Laboratory Methods 2015

MATEC Web of Conferences Volume 89 (2017)

Liberec, Czech Republic
29 - 30 April 2015

Editor:

Petra Dancova

ISBN: 978-1-5108-3365-4

Printed from e-media with permission by:

Curran Associates, Inc.
57 Morehouse Lane
Red Hook, NY 12571

Some format issues inherent in the e-media version may also appear in this print version.

Printed by Curran Associates, Inc. (2017)

For additional information, please contact EDP Sciences – Web of Conferences at the address below.

EDP Sciences – Web of Conferences
17, Avenue du Hoggar
Parc d'Activité de Courtabœuf
BP 112
F-91944 Les Ulis Cedex A
France

Phone: +33 (0) 1 69 18 75 75
Fax: +33 (0) 1 69 28 84 91

contact-edps@webofconferences.org

Additional copies of this publication are available from:

Curran Associates, Inc.
57 Morehouse Lane
Red Hook, NY 12571 USA
Phone: 845-758-0400
Fax: 845-758-2633
Email: curran@proceedings.com
Web: www.proceedings.com

TABLE OF CONTENTS

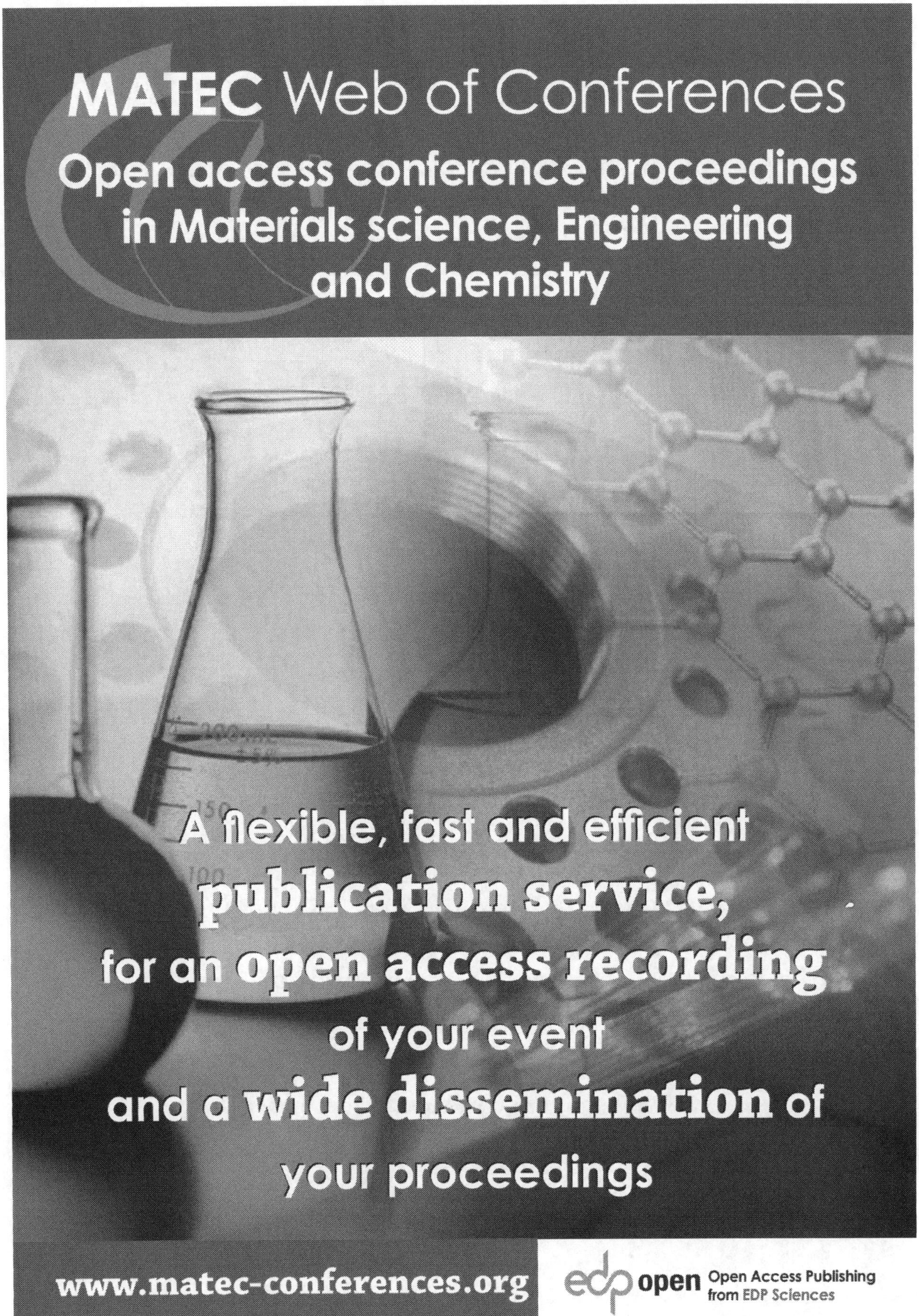
MATEC Web of Conferences
Open access conference proceedings in Materials science, Engineering and Chemistry
A flexible, fast and efficient publication service, for an open access recording of your event and a wide dissemination of your proceedings
www.matec-conferences.org
edp open Open Access Publishing from EDP Sciences

Aims & scope

MATEC Web of Conferences is an open-access journal devoted to the publishing of conference proceedings dealing with Materials science, Engineering and Chemistry.

The journal aims to make conference proceedings visible, citable and widely usable by the scientific community. It offers outstanding services for the publication and dissemination of contents in order to ensure long-term archiving and maximum exposure of scientific events.

Coverage includes all engineering disciplines: civil, naval, mechanical, chemical, and electrical engineering as well as nanotechnology and metrology. The journal also deals with the physical-chemical characterization of materials, their implementation, their resistance in their environment, etc. Others subdisciplines of chemistry, such as analytical chemistry, petrochemistry, organic chemistry, etc., and even pharmacology, are also welcome.

Why publish with MATEC Web of Conferences?

- **Open access:** conference papers available worldwide.
- **Fast publication:** less than 3 months.
- **Total flexibility:** no page restrictions, free unlimited use of colour in the online version, any kind of scientific document welcomed (traditional articles, posters, abstracts and slideshows).
- **High visibility:** all documents indexed in main databases and easily found online.
- **Fair author rights:** CC-BY Creative Commons license.
- **Range of additional publishing options:** printed proceedings, CD-ROMs, copy-editing, etc.

Contact: Isaline Augusto
matec@webofconferences.org

ISSN (electronic edition): 2261-236X

www.matec-conferences.org

edp open Open Access Publishing from EDP Sciences

edp sciences

MATEC Web of Conferences

Statement of Peer review

In submitting conference proceedings to MATEC Web of Conferences, I certify to the Publisher that I adhere to the **Policy on Publishing Integrity** of the journal in order to safeguard good scientific practice in publishing.

1. All articles have been subjected to peer review administered by the proceedings editors.
2. Reviews have been conducted by expert referees, who have been requested to provide unbiased and constructive comments aimed, whenever possible, at improving the work.
3. Proceedings editors have taken all reasonable steps to ensure the quality of the materials they publish and their decision to accept or reject a paper for publication has been based only on the merits of the work and the relevance to the journal.

Title, date and place of the conference

LABORATORY METHODS
JEŠTĚD TOWER – LIBEREC, CZ
APRIL 29-30, 2015

Proceedings editor(s): PETRA DANČOVÁ

Date and editor's signature 22/11/16

Investigation of mixing chamber for experimental FGD reactor

Jan Novosád[1,a], Petra Dančová[1] and Tomáš Vít[1]

[1]Department of Power Engineering Equipment, Faculty of Mechanical Engineering, Technical University of Liberec, Studentska 2, 46117, Liberec, Czech Republic

Abstract. This article deals with numerical investigation of flow and mixing of air and sulphur dioxide SO_2 in designated mixing chamber. The mixing chamber is a part of experimental laboratory reactor designed for simulating the flue gas desulfurization (FGD) process. Aim of this work is the numerical investigation of effect of different mixing chamber geometries to mixture composition, especially to mass fraction of sulphur dioxide. Using of similar concentration of sulphur dioxide in the experimental reactor as in the real process is necessary to be able to make additional research. Conclusion describes the effect of different geometries of mixing chamber to mixing. The aim of this work is to develop perfectly works mixing chamber, which will be manufactured and then implemented into experimental FGD reactor. The results will be validated by experiment after the mixing chamber will be manufactured.

1 Introduction

Coal-fired power plants in the Czech Republic are still a significant source of electricity. They produce a large amount of gases, which are harmful for the environment, especially NO_x and SO_2. Amount of pollutants generated during combustion is several times more than the amount prescribed by legislation. To reduce the amount of SO_2 in the emissions, it is necessary to install the flue gas cleaning device.

The process of desulfurization is very complicated and depends on many parameters. There are lot of chemical reactions, thermal effects and finally the fluid flow and heat transfer. Marocco and Inzoli in work [1] described the mathematical model with all governing equations and source terms. They used them in numerical simulation of a pilot plant open-spray tower.

These towers for FGD are several tens of meters tall and few meters wide cylinders. Processes in spray tower occurring at very small size. Therefore if we want to make numerical simulation, the computational mesh should be very fine, so it consists of several million cells.

Currently there is no computational capacity to make simulation in short time with sufficient accuracy. Therefore it is appropriate to create a scale model of spray tower, where we can simulate the FGD process. This model could be used to verify numerical simulations. The advantages of this equipment is lower prize and huge opportunities for research, because the real spray tower is very needful equipment in flue gas cleaning and there are no possibilities to change regimes or equipment while the power plant is running.

Principles of FGD process is well-known. It is based on reactions of sulphur dioxide with limestone. Products of these reactions are especially gypsum and clean air. A lot of researches are interested in research which consists improving efficiency of FGD, utilization of waste products, savings of chemical agents etc.

The whole process was described by many authors, f.e. by Córdoba in work [3]. Novosád and Vít in work [4] presented the beginning of their research of FGD. They found that they need to make an experimental reactor to verify their numerical results. One of prerequisites for the proper function of experimental reactor is the preparation of flue gas.

Aim of this work is to investigate effect of mixing chamber to mixing SO_2 and air and using the results to design the mixing chamber component.

2 Experimental reactor

The experimental reactor is designated as a scale model of FGD spray tower with some modifications that allows operation under different conditions (regimes) and changings of components.

Block scheme of the reactor is shown in figure 1. In real reactors the source of flue gas is the combustion of coal. To have better control over the concentration the gas cylinder filled with sulphur dioxide is used as a source of flue gas. Gas in the cylinder is very concentrated (over 99.9%). Therefore it is necessary to mix it with air to gain the concentration which is usual in industrial process. This concentration is around 5000 $mg \cdot m^{-3}$ which equals to mass fraction 0.166%.

[a] Corresponding author: jan.novosad@tul.cz

 DOI: 10.1051/matecconf/20178901001

In the reaction zone should be same conditions as in the real reactor. Reaction zone is cuboid with square cross section. Section area of the reaction zone is

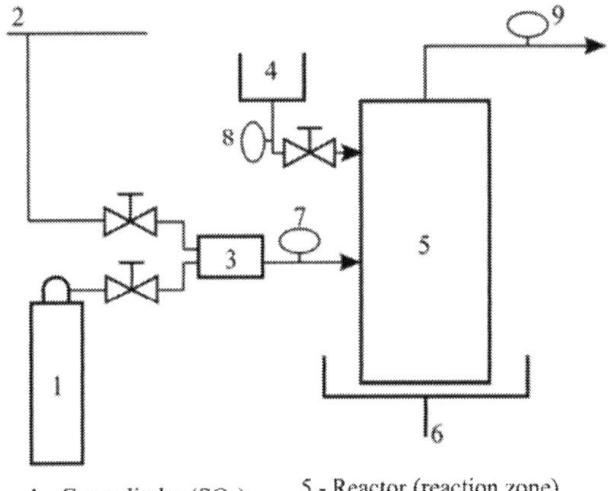

1 - Gas cylinder (SO_2) 5 - Reactor (reaction zone)

2 - Pressurized air 6 - Gypsum reservoir

3 - Mixing chamber 7, 8, 9 – Measuring devices

Figure 1. Scheme of experimental FGD reactor.

S = 0.01 m^2, vertical velocity of gases should be around v = 3 m·s^{-1}. so the mass flux is:

$$\dot{m} = \varrho \cdot S \cdot v, (kg \cdot s^{-1}) \quad (1)$$

where ϱ is the density of mixture (air+SO2).

In our case:

$$\dot{m} = 1.2 \cdot 0.01 \cdot 3 = 0{,}0361\ (kg \cdot s^{-1}) \quad (2)$$

The mass flux of component *i* is defined by:

$$\dot{m}_i = w_i \cdot \dot{m} \quad (kg \cdot s^{-1}) \quad (3)$$

where w_i is the mass fraction of i-th component.

Following equation (3) using mass fractions described above leads to calculations of mass flux of air $m_{air} = 0.036$ kg·s^{-1} and SO_2 $m_{SO2} = 6 \cdot 10^{-5}$ kg·s^{-1}. These fluxes are used as boundary conditions in numerical simulations.

3 Numerical simulation

Numerical investigation was performed in ANSYS Workbench environment, especially modules for creating models (Design Modeler), mesh (Meshing) was used. Ansys Fluent v. 16.0 was used for CFD calculations.
Simulations were performed in three stages. In first stage the 2D mesh quality (element size) impact to results was tested. In the second stage different geometries were calculated using 2D mesh with element size determined in stage one. The last stage include simulations of the better variant evaluated from stage two using 3D mesh.

3.1 Geometry of models

The principle of mixing chamber is based on two inlets for each mixture component and one common outlet. For the investigation several types and variants of geometries were designed (see Figure 2). All geometries have similar dimensions of inlets, outlet, length of the chamber and distance between air and SO_2 inlets.

The first type (see Fig. 2a) have the air inlet in the direction of main cavity, inlet of sulphur dioxide is connected from the side. The effect of the angle α of SO2 inlet tube was investigated. The angle varies in interval of $\alpha = \{15°, 30°, 45°, 60°, 75°, 90°\}$. Special variant of this geometry is the tube (nozzle) inserted to the axis of main cavity (see Fig. 2b).

The second one have SO_2 inlet in the main direction and two (symmetric) air inlets from sides (see Fig. 2c). The effect of the angle α of air inlet tubes was investigated. The angle varies in interval of $\alpha = \{15°, 30°, 45°\}$.

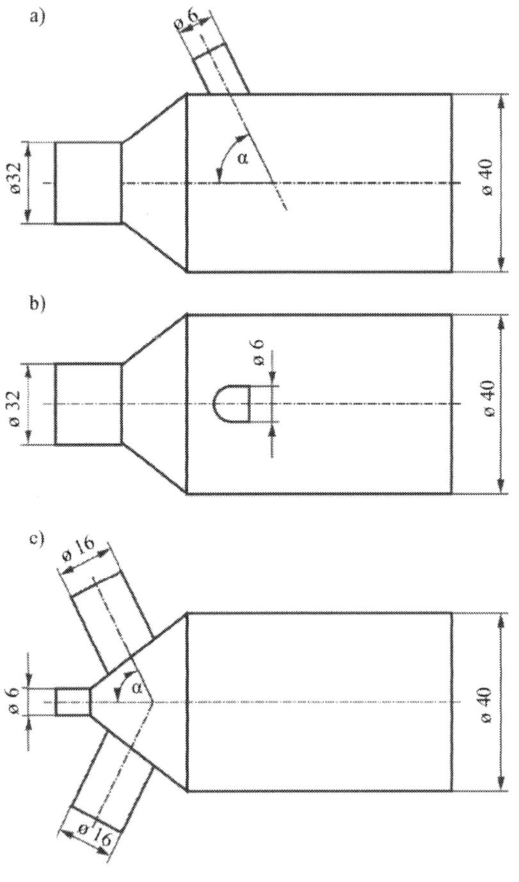

Figure 2. Sketches of model types and variants.

3.2 Mesh

Meshes for all models was created using ANSYS Meshing. 2D meshes consists of quad elements with different size varies of {0.2, 0.5, 0.7, 1, 2} (mm). Labels of variants for 2D follow the scheme: „2D-el. size-angle_α". Variant shown in figure 2b) is labelled as „2D-el. size-AX." Examples of meshes are shown in figure 3.

 DOI: 10.1051/matecconf/20178901001

After the evaluation of first stage of calculations, the suitable element size (0.7 mm) were determined and this element size was used for additional stages.

For 3D mesh hexahedral elements were used. Labels of variants for 3D follow the scheme: „3D-angle_α.“

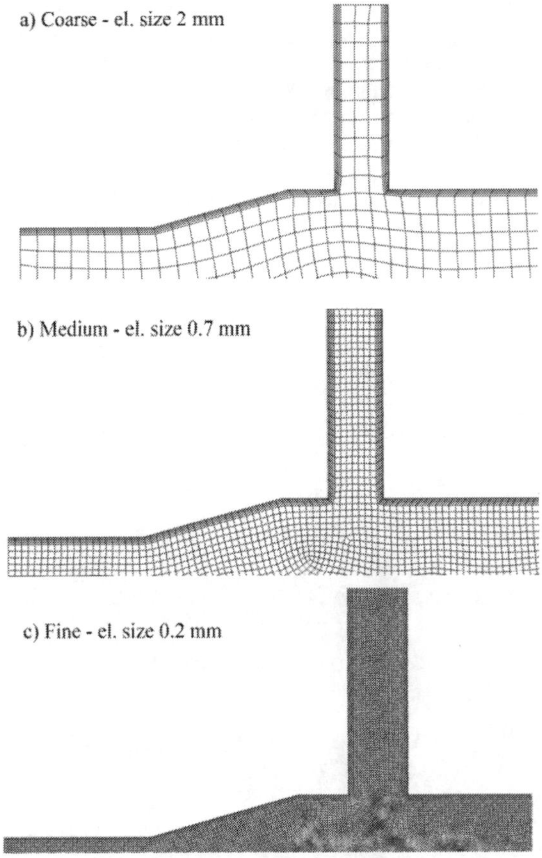

Figure 3. 2D meshes with different element sizes.

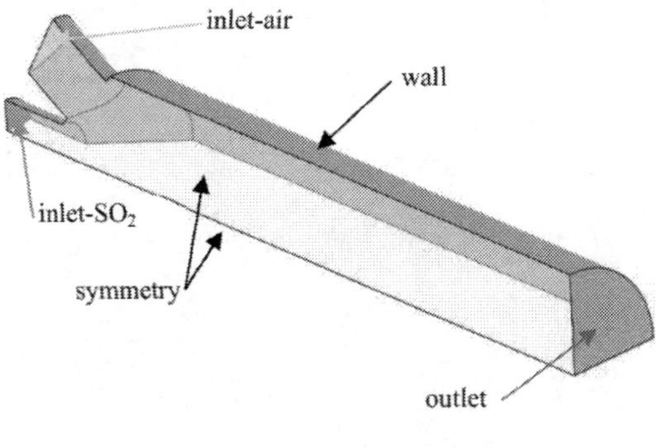

Figure 4. Boundary conditions – 3D model.

3.3 Boundary conditions

At the inlets of air and SO_2 were assigned *mass-flow-inlet* boundary condition with mass flow value specified in section 2. Inlets of air are tubes with diameter of 32 mm (see Fig. 2a, b) or 2 x 16 mm (see Fig. 2c). Inlets of SO_2 are tubes with diameter of 6 mm in all cases.

At the outlet (tube with diameter of 40 mm) were set *pressure-outlet*.

To simplify the 3D model *symmetry* boundary conditions were used. 3D model with labelled boundary conditions is shown in figure 4.

3.4 Solution

At the inlets of air and SO_2 were assigned *mass-flow-inlet* boundary condition with mass flow value specified in section 2.

Flue gas was considered as ideal gas mixture of air and SO_2. The energy equation was set to on. Viscous model SST k-ω was used. We used steady flow and *pressure based* solver. Species transport model with no reactions was used.

3.5 Methodology of the results assessment

The results obtained by numerical simulation were mass fraction and velocity fields. Values at inlet and outlet boundary conditions were evaluated using *Area-weighted-average* method. To plot the mass fraction across the cross section, additional surfaces and lines were created.

To compare different geometries we use the mass fraction of SO_2 to rate the desired function of mixing chamber, especially to gain desired mass fraction on outlet and its uniformly distribution in fluid flow.
Turbulent kinetic energy fields is displayed to show the efficiency of mixing and places where the mixing should be increasing by geometry modification.

Labels of variants for 2D follow the scheme of file names: „2D-el. size-angle“. For 3D the file names are in this way: „3D-angle“.

4 Results

Results are divided into three sections accordingly to the stages of numerical simulations. Labels of variants for 2D follow the scheme of file names: „2D-el. size-angle“. For 3D the file names are in this way: „3D-angle“

4.1 Mesh quality impact

The mesh quality impact was investigated on 2D variant shown in figure 2a) with angle of SO_2 inlet tube $\alpha = 90°$ and boundary conditions described in section 3.3 were used.

Profiles of mass fraction of SO_2 across the outlet area is shown in figure 5, detail of profile near the upper wall

DOI: 10.1051/matecconf/20178901001

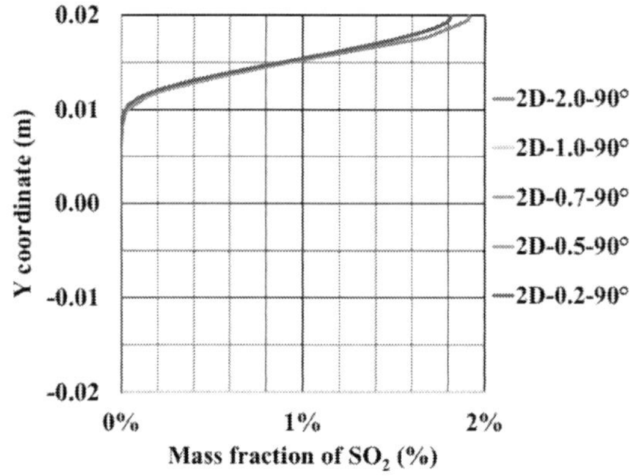

Figure 5. Mesh quality impact.

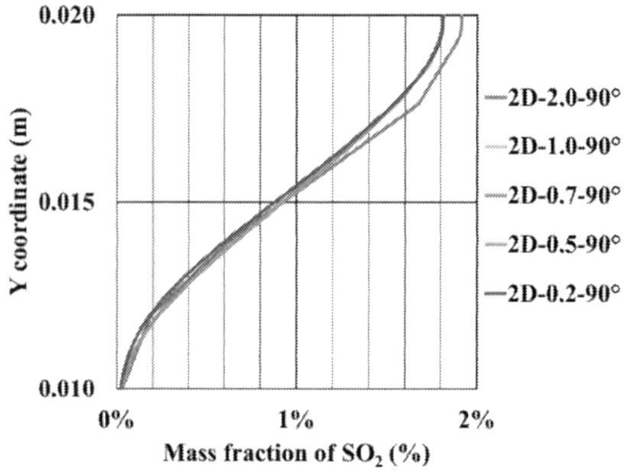

Figure 6. Mesh quality impact - near wall detail.

is shown in figure 6. Coordinate Y = 0 (m) means the point on the tube axis.

It was founded that the coarse mesh (element size 0.2 mm) causes higher mass fraction of SO_2 than in other cases. Detailed graph (see Fig. 6) shows, that the curves for all cases have similar course. It can be said that curves for element size in the middle of the range are closer.

This leads to the conclusion that it is suitable to use mesh with element size in interval {0.5, 1.0} (mm). For additional cases element size 0.7 mm was chosen.

4.2 Effect of SO_2 tube angle

The effect of the angle of SO_2 inlet tube was investigated on 2D variant shown in figure 2a) with angle of SO_2 inlet tube varies in $\alpha = \{15°, 30°, 45°, 60°, 75°, 90°\}$. Additional simulated variant was the geometry shown in figure 2b). Mesh element size was set to 0.7 mm and boundary conditions described in section 3.3 were used.

Profiles of mass fraction of SO_2 across the outlet area for different variants are shown in figure 7. Coordinate Y = 0 (m) means the point on the tube axis.

For individual variations it is evident that the mass fraction of SO_2 depends insignificantly on the angle of SO_2 inlet tube. To change the mass fraction profile across the outlet area, we need to change the geometrical arrangement (see Fig. 7, variant 2D-0.7-AX) to position the inlet of SO_2 into the axis of main tube. Examples of spread of SO_2 for two variants are shown in figure 8 using the SO_2 enthalpy field, where the way of SO_2 could be seen.

4.3 Effect of air tube angle

Following previous results the inlet of SO_2 should be arranged into the axis of main tube. To provide simpler construction of mixing chamber the geometry concept follows figure 2c), where are two inlets of air on both sides of mixing zone. The effect of the angle of air inlets was investigated with angle of inlet tube varies in $\alpha = \{15°, 30°, 45°\}$.

Velocity fields of X-velocity (in the direction of main

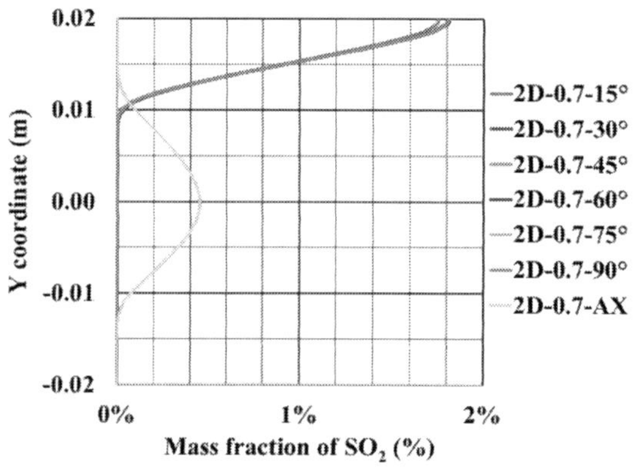

Figure 7. Effect of SO_2 inlet tube angle.

a) 2D-0.7-45°

b) 2D-0.7-AX

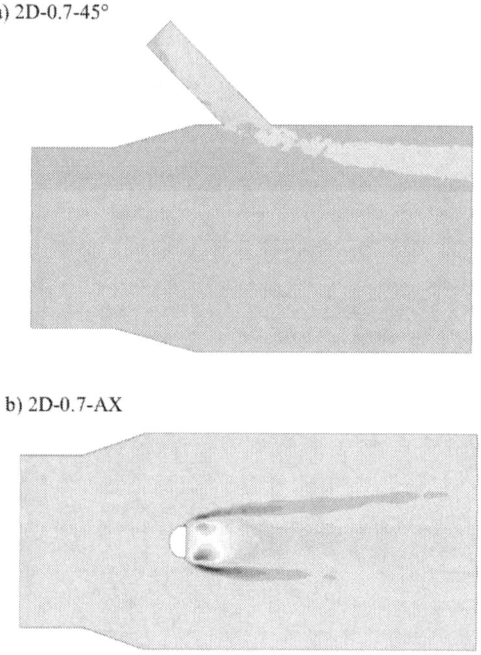

Figure 8. Spread of SO2 - examples.

 DOI: 10.1051/matecconf/20178901001

tube axis) for all cases on symmetry planes are shown in figure 9. This figure shows mirrored model, calculations were performed for the quarter of displayed model. In figure 9 can be observed that with increasing the angle of inlet air tube the zone with high velocity appears on the axis. With higher angle this zone is placed closer to SO_2 inlet, which could cause the blockage in the sense of less amount of SO_2, which can flow in the X direction, so it is deflected to the side of mixing chamber. This deflection causes more time spending in mixing chamber.

Profiles of mass fraction of SO_2 in the outlet (see Figure 10) were evaluated in two directions of model [coordinate system – axis Y (vertical) and axis Z (horizontal)]. First figure for mass fraction in Y direction shows that the most uniform profile of mass fraction is for air inlet tube angle $\alpha = 15°$. In Z direction is the situation exactly the opposite, the most uniform profile is for tube angle $\alpha = 45°$.

It can be assumed that the values of mass fraction are distributed between Y and Z axis. Therefore simply averaging of values in Y and Z direction was carried out. The result is shown in figure 11. From this figure we can see that the most uniform profile is for angle $\alpha = 45°$.

The best variant for application to design of mixing chamber could be illustrated by streamlines. Figure 12 shows streamlines begins at inlets of air (red lines) and SO_2 (green lines). The figure illustrates higher mixing for the case with $\alpha = 45°$. In the zone where two fluid flows are interacting, the SO_2 stream deflects and is brought into rotation which helps to mixing.

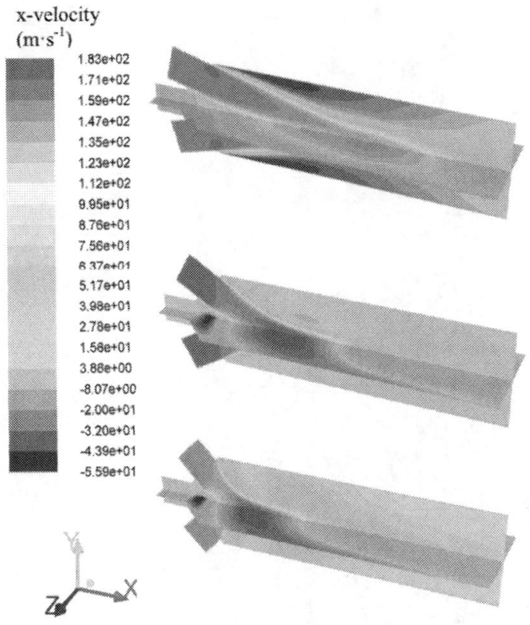

Figure 9. Contours of x-velocity for $\alpha = 15°, 30°, 45°$. (from the top to bottom)

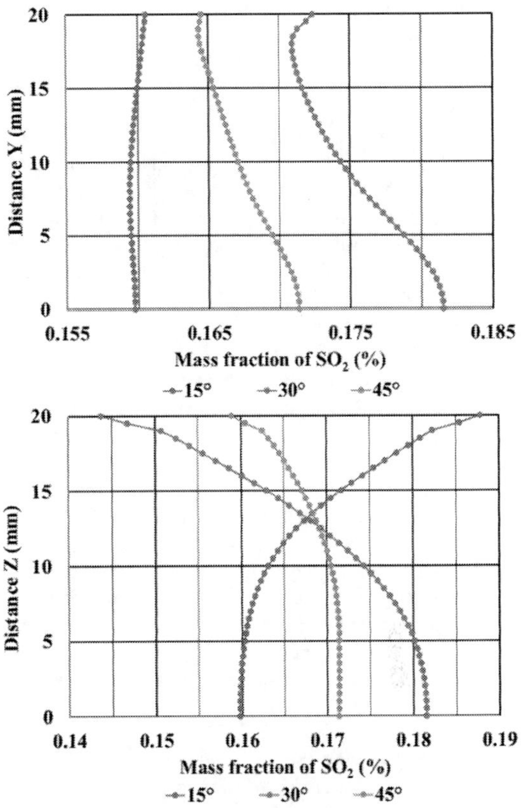

Figure 10. Mass fraction profile on outlet.

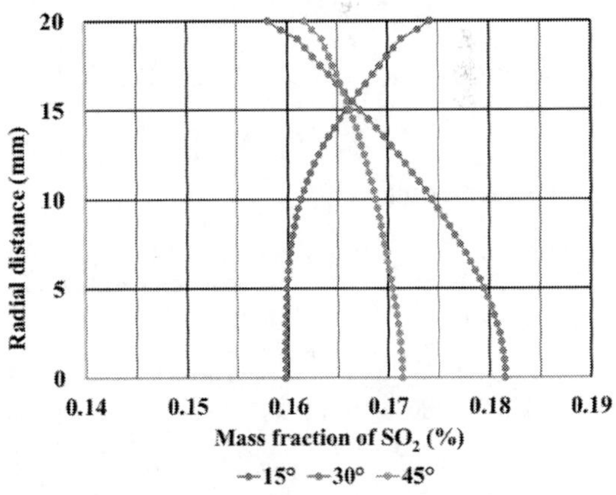

Figure 11. Averaged mass fraction profile on outlet.

DOI: 10.1051/matecconf/20178901001

5 Conclusion

Numerical simulations performed in this work do not reflect the kinetics of chemical reactions. Simulations of various geometries of mixing chamber for mixing air and SO_2 in three stages were performed. The calculations give detailed information about the distribution of the SO_2 across mixing chamber outlet area. The results of each stage were used for next stage to gain the useful geometry for mixing chamber at the end of investigation.

During 2D simualtions for varies geometries was founded that the inlet of SO_2 from the side of mixing chamber is not useful, because it causes high mass fraction of SO_2 near the wall, but the zone around axis is not fullfilled. Better solution is to insert the SO_2 inlet into the axis of mixing chamber.

In last stage 3D models of mixing chamber were simulated. The airflow was divided into two streams inputted by two inlets positioned on the sides of mixing chamber under angle α. This embodiment is suitable for mixing because it allows uniform profile of SO_2 concentration at the outlet. As the best option seems to be variant with air inlet tube angle $\alpha = 45°$. Maybe the additional research should be carried out to investigate the effect of more angles near to 45°.

The future work consider the construction design of whole mixing chamber including connection for hoses and pipes. After it will be manufactured the control measurement of SO_2 fraction and velocity profiles will be performed. After that the mixing chamber becomes a part of experimental FGD reactor.

Figure 12. Streamlines of air and SO_2 for $\alpha = 45°$.

Acknowledgement

Author gratefully acknowledges financial support by MŠMT under the project CZ.1.07/2.2.00/28.0311 and project SGS 21000.

References

1. L. Marocco, F. Inzoli. Multiphase Euler–Lagrange CFD simulation applied to Wet Flue Gas Desulphurisation technology. International Journal of Multiphase Flow, **35**, 2009.
2. Ansys Inc., Fluent User´s Guide Release 16, 2015.
3. P. Córdoba. Status of Flue Gas Desulphurisation (FGD) systems from coal-fired power plants: Overview of the physic-chemical control processes of wet limestone FGDs. Fuel, Vol. **144**, 2015.
4. J. Novosád, T. Vít. Numerical simulation of flow in the wet scrubber for desulfurization. Experimental Fluid Mechanics 2013, November 18.-21., 2014.

MATEC Web of Conferences **89**, 01002 (2017)
Laboratory Methods
DOI: 10.1051/matecconf/20178901002

Experimental analysis of the sewing machine

Jaroslav Beran[1], Martin Bílek[1], Jiří Komárek[1a], and Pavel Němeček[2]

[1]Technical University of Liberec, Dept. of Textile Machine Design, Studentská 2, 461 17 Liberec 1, Czech Rep.
[2]Technical University of Liberec, Dept. of Vehicles and Engines, Studentská 2, 461 17 Liberec 1, Czech Rep.

Abstract. This paper is concerned with experimental analysis of the behaviour of mechanisms of sewing machines. There has been carried out mapping of the sound field in the vicinity of the sewing machine, in order to locate the source of impact loading and to establish the overall hygienic loading of the machine. Based upon this analysis, the acceleration of the needle bars has been measured. These two measuring have led to the conclusion that it is the cam mechanism which most contributes to the noise and impact loading.

1 Introduction

The present trend in the development of sewing machines is to shorten the sewing times in the sewing process and to increase the productivity of this process. Other requests upon sewing machines manufactured nowadays are e.g. silent running, minimum vibrations, long service life of employed mechanisms and easy operation of the machine [1-3]. The analysed Deco 2000 Hand Stitch sewing machine uses the floating needle system, developed by AMF Reece, to produce a "true hand stitch" as represented in figure 1.

Figure 1. Hand stitch types.[7]

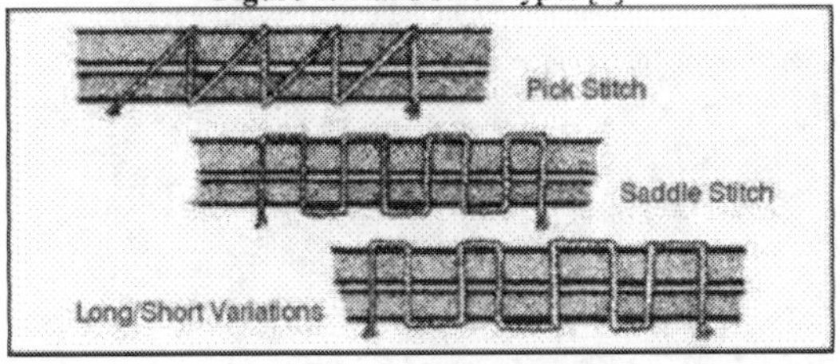

The machine uses a double pointed needle which has an eye in the middle, as represented in figure 2. By the use of two needle bars, one above and one below the work plate, a length of thread is passed through the material on every stitch, reproducing the seamstress work, accurately and at an incomparable speed. The needle bar is represented in figure 3.

Figure 2. Double pointed needle. .[7]

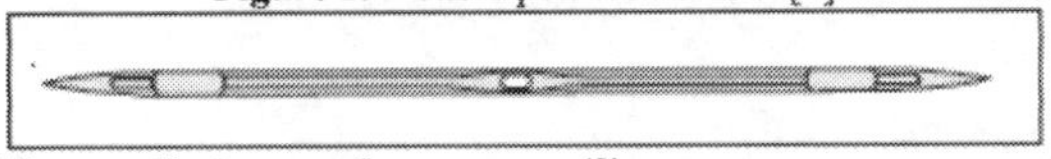

The needle bar performs a rectilinear reverse movement obtained by a cam mechanism.

[a] Corresponding author: jiri.komarek@tul.cz

Figure 3. Needle bar. .[7]

The sewing machine is operated at 190 rpm now. The aim of the manufacturer is to operate it at 250 rpm. This velocity can be achieved without any alterations of the machine design; however, in such a regime the sewing machine shows considerable vibrations and high levels of intensity of noise, which are undesirable, and because of hygienic regulations, not admissible legally. At the velocity of 250 rpm, the machine components wear out more, too, with shorter service life of the machine as consequence. The assignment of the experimental analysis has been to establish the hygienic loading of the machine with respect to its vicinity, and to find out critical points of the machine. For this purpose, the intensity of the sound has been measured, and the resultant weighted noise level has been determined. As the next step, there have been measured the values of acceleration of the needle bar in dependence upon the angular displacement of the main cam. The measuring points have been chosen on the needle bar. The measuring has been carried out both on the upper needle bar and on the lower one [4-6].

2 Mapping of sound field in the vicinity of the sewing machine

The measuring of sound field has been performed by means of equipment of the firm Brüel & Kjaer. There has been employed a dual-channel analyser, a transducer of intensity of the sound, an acoustic calibrator and a sound intensity calibrating set. The measuring equipment has been set up to measure the intensity of sound, and adjusted by means of the calibrating set. The sewing machine has been surrounded with gauging grid with partial areas 10 cm x 10 cm. the measuring areas have been set out in the front and in the rear of the machine, in its L.H. and R.H. sides and in the upper section of the machine. The measured object has been operated at 190 rpm and 250 rpm in uncovered state, and in the operating regime of 250 rpm in covered state. In the centres of partial areas of the grid, there have been measured both the level of active intensity of sound (the vector component in the direction off the source) and the level of acoustic pressure. By means of the software, in the gauging grid there have been established the areas with a constant range of sound intensity. These areas have been visualised on the background of a photo of the measured object, for the purpose of the overall weighted noise level. The figure 4 shows the spectrum of the sound intensity level and of the level of acoustic pressure. The spectrum has been established at the front side (at the attendant's place) in the centre of the segment of the gauging area tracing out the position of the visible section of the sewing machine.

From the spectrum, there can be seen the dominant amplitude at the frequency 40Hz. Moreover, high amplitude can be seen at the frequency 315Hz. There is an increased level in the frequency band around 1-2 kHz, too. The frequency spectrum has a comb-like shape, which points to generation of impacts in the machine. Once the machine has been equipped with covers, the peaks in the frequency spectrum diminish. The difference between the level of sound intensity and the level of acoustic noise is due to the diffusion (reflectance) of the sound field.

The following table shows the overall weighted levels of sound powers established in individual gauging areas.

Table 1. Levels of sound powers (dB).

	without covers		with covers
rpm	250	190	250
in the front	83,4	80,1	83,6
in the rear	80,8	77,7	78,2
R.H. side	78,0	74,3	78,2
L.H. side	76,8	73,4	75,7
upper section	80,0	76,6	79,7

Figure 4. Example of the spectrum of sound intensity level and level of acoustic pressure.

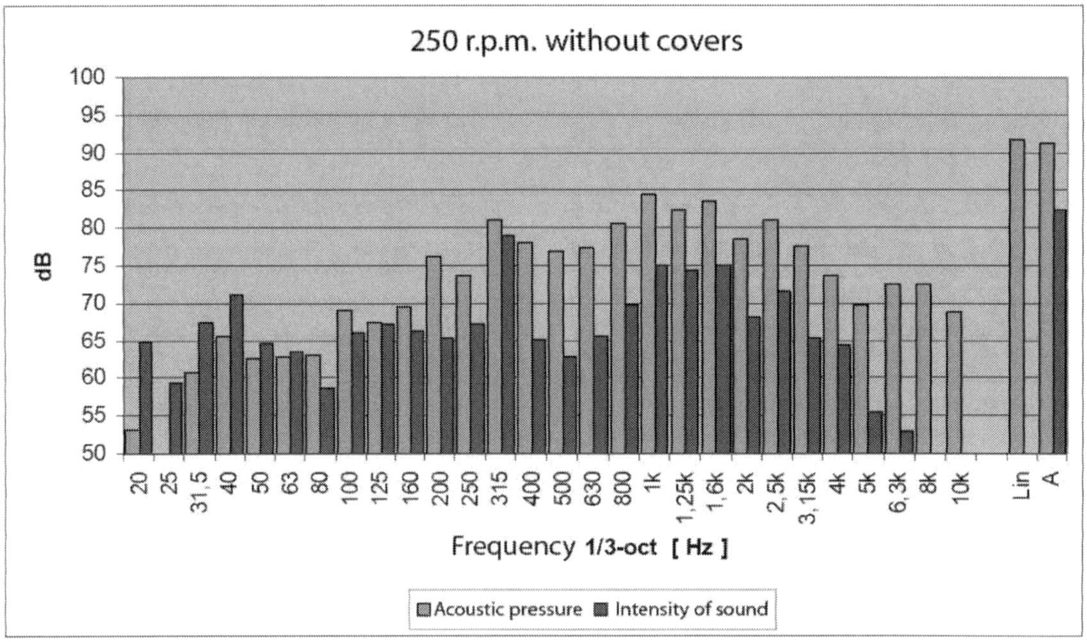

Figure 5. Example of distribution of the sound intensity level (front side at 250 rpm, with covers)

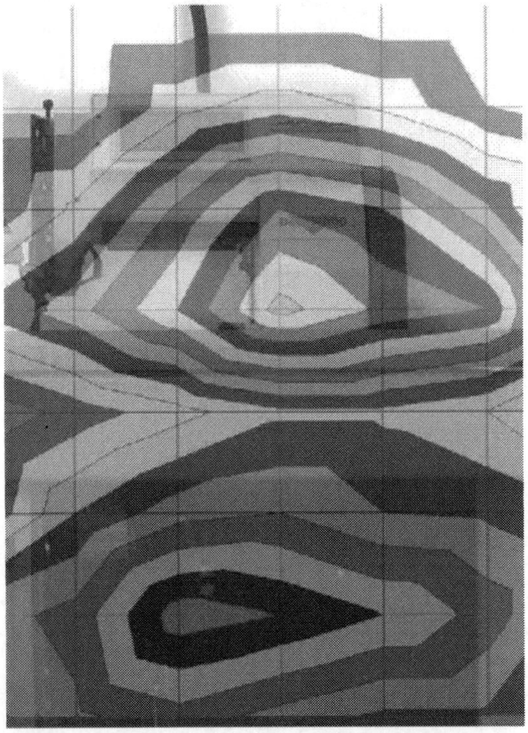

Figure 6. Example of distribution of the sound intensity level (right side – at 250 rpm, with covers)

From this table there is evident that the highest sound power is emitted to the front side, i.e. towards the machine attendant. Moreover, we can state higher values of sound power in the covered state compared to the uncovered one in the front and in the R.H. side. Consequently, the employed covers have not been devised adequately, as far as acoustic loading of the machine is concerned.

In the figs. 5 and 6, we can observe an example of distribution of the sound field in the front side and in the R.H. side at the velocity 250 rpm in covered state. By means of these two figures, we can locate the dominant source of noise and of impact loading, namely the area of the contact of radial cam with the roller, connected to the double-arm lever. This cam mechanism provides for the movement of the two needle bars from their upper position to the lower one and back. This analysis and the subsequent localisation of the biggest source of the noise have induced us to measure the acceleration of the needle bars

3 Measuring and evaluation of the acceleration of needle mechanism

For sensing the values of the acceleration, there has been employed a piezoelectric acceleration sensor of the type 4520 with a charge amplifier of the firm Brüel & Kjaer. For sensing the angular displacement of the cam, there has been employed an incremental sensor marked IRC 305 of the firm LARM a.s. (Inc.). For processing the ascertained values there has been employed the centre MGCXplus of the firm HBM. The measuring of acceleration of the needle bars has been performed both at the velocities 190 rpm and 250 rpm The ascertained courses of the acceleration have been evaluated in dependence upon the angular displacement of the main cam. The ascertained courses have been hardware filtered by a high-pass filter, on which the filtration values 1000 Hz and 40 Hz have been set up. This filtration of the input signal has brought about flattening of the log, thanks to elimination of high-frequency effects. The log processed in this manner is more transparent and more suitable for its further analysis. If a filter of the value 40 Hz is set up, we obtain a theoretical course of the acceleration nearly, generated by the link mechanism driven by a cam.

The tables 2 – 4 show the maximum and minimum values of acceleration of needle bars for individual velocities and for different settings of the filter.

From the recorded courses of acceleration it is obvious unequivocally that needle bars are exposed to impact loading. These impact loads are generated in particular by the plays in the entire cinematic chain (play in the cam and link mechanism, plays in the seating of needle bars in the machine frame etc.). An incorrectly devised lift dependence on the cam can contribute to impact loading as well.

The acceleration occurring on the upper needle bar is markedly higher than that on the lower needle bar, which can be caused by different adjustment of the needle bar mechanisms. There has been analysed the original adjustment of the machine from the manufacturing plant.

The maximum and minimum acceleration appearing on the measured components is many times higher than the theoretical values (these have been found out from the supplied data of geometry of the driving cam).

4 Conclusion

There has been carried out measuring of the sound intensity. This measuring has determined the overall weighted level of sound powers ascertained in the gauging areas. This measuring also indicates the maximum value of the level of sound power 83.6 dB at the velocity 250 rpm. in the covered state. Furthermore, the measuring has located the area with the biggest source of noise; it has been in particular the area of contact of the roller and the cam. In the second part of the experimental analysis, there has been performed measuring of acceleration of the needle bar. This measuring has established the maximum and minimum values of acceleration of the said components, and has identified the impact loading of the needle mechanism.

Figure 7. Example of the courses of acceleration of upper and lower needle bars

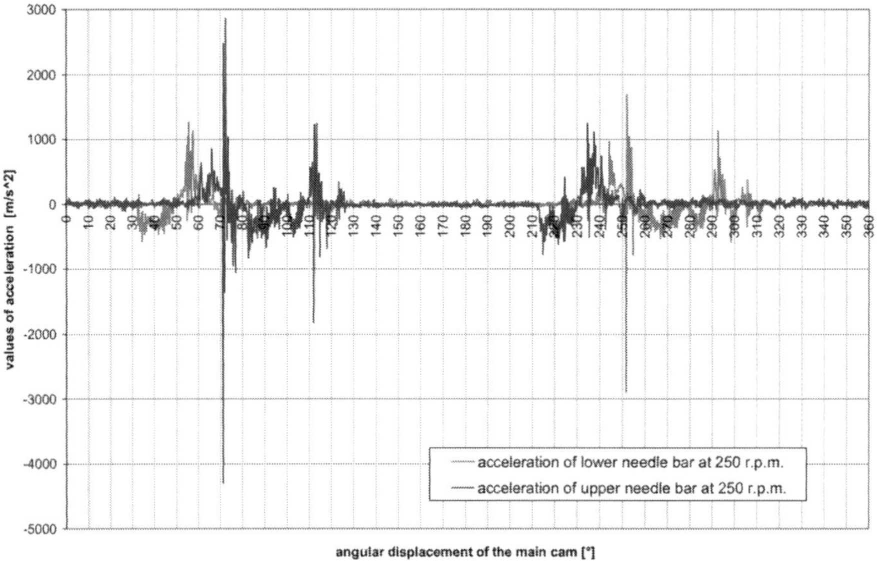

Figure 8. Course of acceleration of the needle bar if employing high-pass filters, compared with theoretical

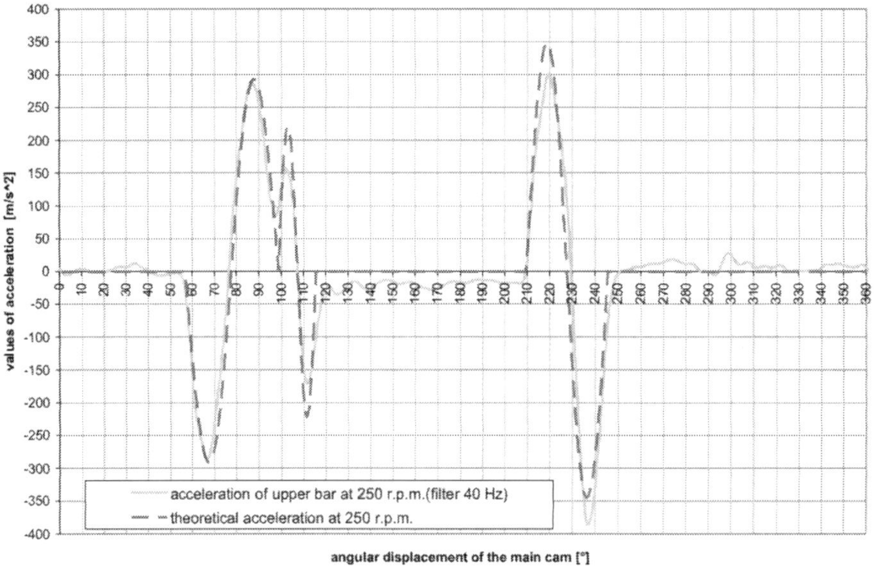

Table 2. Maximum and minimum values of acceleration without employing a filter.

Without filter		Acceleration of lower needle bar ($m.s^{-1}$)	Acceleration of upper needle bar ($m.s^{-1}$)
Minimum value	190 rpm	- 1762.9	- 4545.1
	250 rpm	- 3305.6	- 5322,7
Maximum value	190 rpm	1742.1	2863.1
	250 rpm	1965.0	3407.2

Table 3. Maximum and minimum values of acceleration employing filter 1000 Hz .

Filter 1000 Hz		Acceleration of lower needle bar ($m.s^{-1}$)	Acceleration of upper needle bar ($m.s^{-1}$)
Minimum value	190 rpm	- 767.0	- 1833.7
	250 rpm	- 1510.7	- 2288,7
Maximum value	190 rpm	500.0	2258.8
	250 rpm	757.7	2627.1

Table 4. Maximum and minimum values of acceleration employing filter 40 Hz .

Filter 40 Hz		Acceleration of lower needle bar ($m.s^{-1}$)	Acceleration of upper needle bar ($m.s^{-1}$)
Minimum value	190 rpm	- 243.0	- 254.4
	250 rpm	- 347.7	- 386.2
Maximum value	190 rpm	193.9	208.1
	250 rpm	289.5	304.0

Acknowledgement: The paper has been elaborated with financial support of TUL in the framework of specific university research competition.

References

1. Komárek, J., Beran, J., Lima, M. Machado, J., Silva, J.: Finding the Optimal Setting of the Sewing Needle Transfer Mechanism Using Simulation Software, The 41st Textile Research Symposium, TRS2012, Guimaraes-Portugal, pp.: 401-404, ISBN 978-972-8063-67-2.
2. Komarek, J.:Influence ot the rubber pad stiffness on the noise of the needle transfer mechanism. IV Miedzynarodowa Konferencja Mlodych Naukovcow Szkol Wyzszych Euroregionu Nysa, Jelena Gora, 2010, pp.:135-139, ISBN 978-83-61719-86-1
3. Beran, J. Pejchar, K., Komárek, J.: Analysis and optimisation of the needle transfer mechanism. ACC JOURNAL XVI 2010, Liberec, 2010, pp.: 5-13, ISSN 1803-9782.
4. Beran, J. Pejchar, K.: Dynamic Analysis of Needle Bar Mechanism of Sewing Machines. Acc Journal XVI 2011, Liberec, 2011, pp.: 60-71, ISSN 1803-9782.
5. Pejchar, K.: Verification of the Mathematical Model of the Needle Bar Mechanism. V. Miedzynarodowa Konferencja Mlodych Naukovcow Szkol Wyzszych Euroregionu Nysa. Jelena Gora, 2011, pp.: 136-143, ISBN 978-83-62708-21-5.
6. Pejchar, K.: Dynamic Analysis of the Needle Mechanism. 18th International Conference STRUTEX, Liberec 2011, Czech Republic, 2011, ISBN 978-80-7372-786-4.
7. DECO 2000 - Decorative Hand Stitching Machine. AMF Reece. [vid. 2015-04-07] : http://www.amfreece.cz/d364-deco-2000.html/

 DOI: 10.1051/matecconf/20178901003

Measuring of beat up force on weaving machines

Martin Bílek[1a], Eva Moučková[2], Josef Skřivánek [1] and Petr Tumajer [2]

[1]Technical University of Liberec, Dept. of Textile Machine Design, Studentská 2, 461 17 Liberec 1, Czech Rep.
[2]Technical University of Liberec, Department of Textile Technologies, Studentská 2, 461 17 Liberec 1, Czech Rep.

Abstract. The textile material (warp) is stressed cyclically at a relative high frequency during the weaving process. Therefore, the special measuring device for analysis of beat up force in the textile material during the weaving process, has been devised in the Weaving Laboratory of the TUL. This paper includes a description of this measuring device. The experimental part includes measurements results for various materials (PES and VS) and various warp thread densities of the produced fabric.

1 Introduction

Fabric in a weaving machine is formed in a discontinuous process based on a cyclic repetition of individual phases of the weaving cycle: opening a shed, interlacing the weft, beat up the weft and closing the shed. This interlaces two mutually perpendicular yarn systems (warp and weft) [5].
Opening a shed and applying the beat up force results in the warp thread being stretched. Because of the high weaving frequencies of modern machines, the warp threads are subjected to cyclic stressing of relatively high frequency [4].

2 Beat up process and its force quantities

A loom is capable of producing fabrics of different weft thread density, i.e. fabrics with varying number of weft threads per centimetre of their length. The number of wefts weaved per unit of time depends on the weaving frequency (machine speed), and the required weft thread density is set by the speed of the fabric being taken up. The fabric is taken up by rough rollers powered by a separate motor. Modern looms allow information concerning the required weft thread density to be uploaded to the control computer which will then set the required take-up speed for the specified loom speed (revolutions of the motor driving the rough rollers) to achieve the required weft thread density [5]. Thus the required weft thread density is set exclusively on the basis of geometric parameters.
The resultant fabric structure including its weft thread density is formed during the beat up process. During this process a mutual force interaction takes place between the weft, the warp system and the fabric. When making a fabric with higher density weft threads, the warp threads due to the beat up stretch and the fabric is "shortened" by value X, i.e. the beat up results in an increase of the tensile force in the warp threads Q_1 and a decrease of the tensile strength in the fabric Q_2. [6]
The force acting on the fabric face at beat up, i.e. the so called beat up force F_b, can then be determined as a difference between the tensile force in the warp threads Q1 and the tensile force in the fabric Q_2:

$$F_P = Q_1 - Q_2 \qquad (1)$$

The beat up force induces a reaction in the fabric, known as weaving resistance R:

$$R = F_P \qquad (2)$$

Figure 1: Textile weaving phases

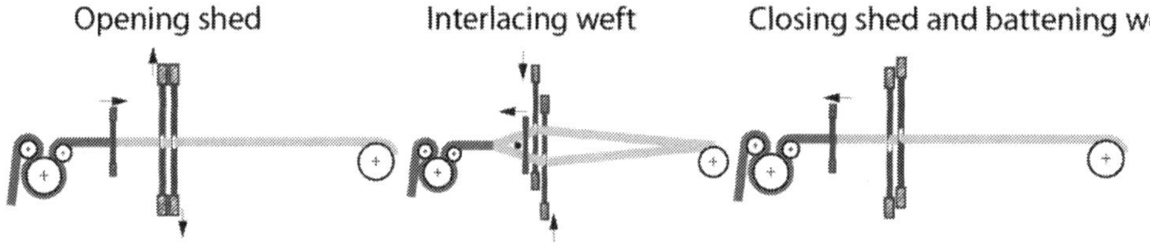

[a] Corresponding author: martin.bilek@tul.cz

Figure 2: Example of tensile forces in the warp as a function of time during the weaving process (machine speed: 500 rpm)

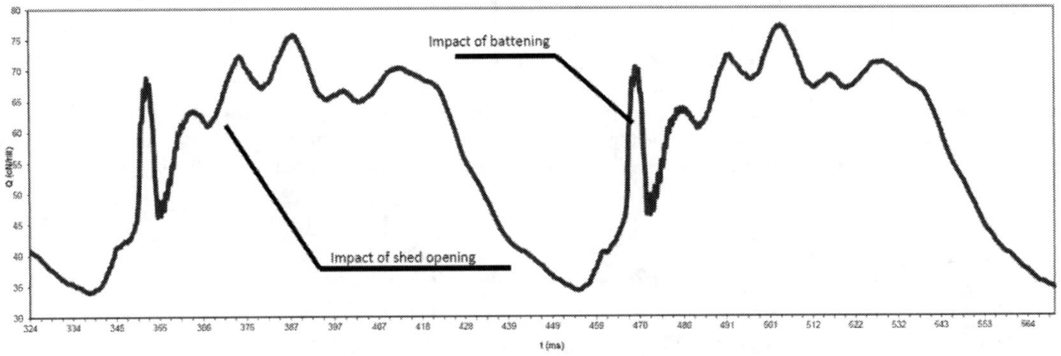

Figure 3: Loom fabric taking-up and setting weft thread density

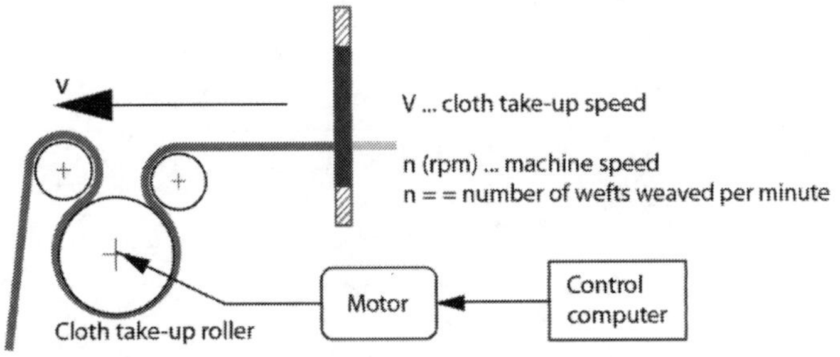

Figure 4: Situation at beat up

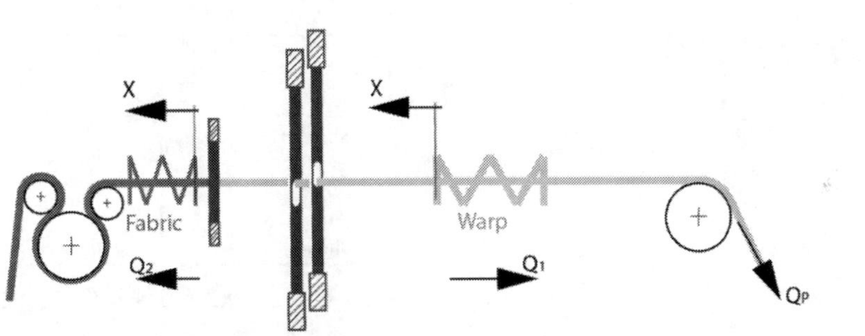

X ... stretching of the warp threads and shortening of the fabric, Q_2 ... tensile force in the fabric, Q_1 ... tensile force in the warp, Q_P ... prestressing set by the let-of motion

Therefore the beat up force and the weaving resistance can be determined as a difference between the tensile force in the warp threads and that in the fabric during the beat up process. This article describes experimental measurements of these quantities. Presented are the results of measurements for various warp thread densities of produced fabrics, as well as various materials being used.

3 Beat up force measuring device

The tensile force in the warp threads and in the fabric was simultaneously measured on a Gamma-8-R rapier loom (manufacturer: Picanol) using strain gauges. The signal received from these stain gauges was digitised and then processed on Waweon equipment [1, 8].

Tensile force in the warp threads was measured by a standard strain gauge sensor which was for this purpose designed by VÚTS Liberec. This sensor was inserted into the warp thread system between the warp beam and heddles.

Tensile force in the fabric was measured by a specially designed sensor. A hard aluminium beam was placed across the entire width of the weaving machine. A strain gauge was positioned between the beam and the fabric, whose one end was fastened to the beam and a "scanning rod" was attached to the other end. This rod was pushed into the fabric (the size of the scanning rod girding determines the equipment's sensitivity). To ensure that the breaking of the fabric in the vicinity of the point of measurement was equal, dimension-wise identical jigs had to be located on the right and the left hand side of the strain gauge.

Both sensors were calibrated to ensure that the value of the output signal determined tensile force in cN in a single warp thread. Tensile forces in the warp threads and in the fabric as a function of time can be illustrated in the Waweon equipment, and these values can then be exported to text files. Further processing is carried out in an Excel spread sheet.

Figure 5: Positioning of sensors and detectors on the loom

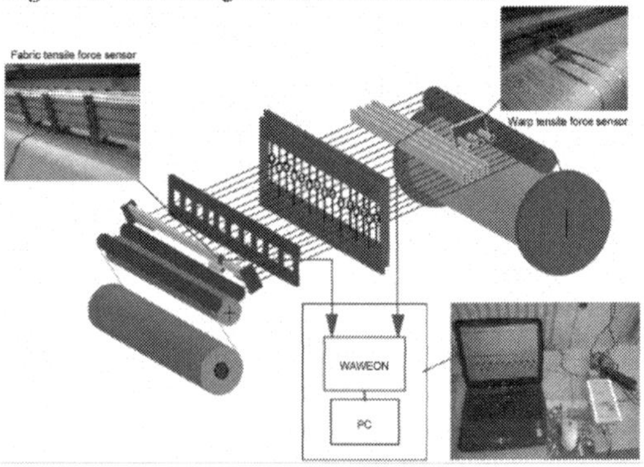

Figure 6: Fabric tensile force sensor

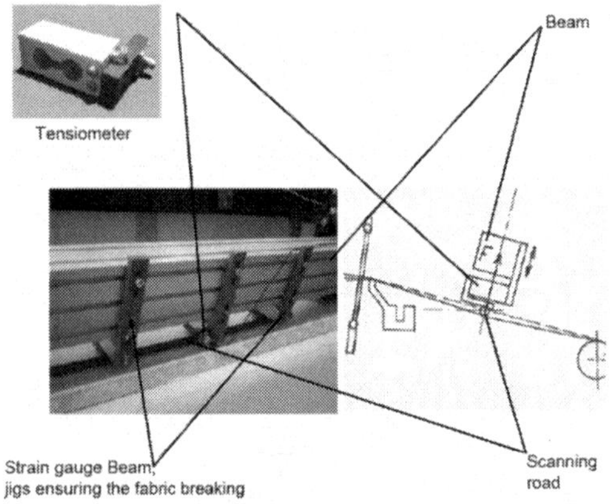

4 Example of measurement results

The results of weaving resistance measurements for various materials (PES and VS) and various warp thread densities of the produced fabrics (from 12 threads per cm to 22 threads per cm) are presented in this article. The measurements were performed with the machine running at a stabilised speed of 300 rpm.

The figure below is an example of the results of measurements taken at weft density 18 threads/cm and material PES. Illustrated in a common coordinate system are tensile forces in warps Q_1 and tensile forces in the fabric Q_2. Presented time interval contains 4 weaving cycles.

Subsequently using an Excel spread sheet, the difference between the tensile force in warps Q_1 and in the fabric Q_2 can be expressed, where the difference represents the beat up force F_p and weaving resistance R, respectively. The weaving resistance as a function of time is illustrated in the figure no 8:

Furthermore, readings of values of maximum weaving resistance were taken for 8 beat up cycles, and these values were statistically processed. The values of maximum weaving resistance R_{max}, standard deviation and a 95% confidence interval for all performed measurements are presented in the table below.

Table 1. Measurement conditions

Speed (rpm)	300	
Reed width (cm)	178	
Warp density (threads/cm)	16	
Fabric weave	Plane	
Material in warp and weft	2 x 25 tex 100% POP	2 x 25 tex 100% VS
Let off motion (cN/thread)	53	42
Weft density (threads/cm)	12, 14, 16, 18, 20, 22	12, 14, 16, 18, 20, 22

Table 2. The values of maximum weaving resistance

Material: 2 x 25 tex 100% PES						
Weft density (threads/cm)	12	14	16	18	20	22
Rmax (cN/thread)	13	22	31	40	52	55
Standard deviation	3	2	3	2	2	4
95% confidence interval	2	1	2	1	1	3

Table 3. The values of maximum weaving resistance

Material: 2 x 25 tex 100% VS						
Weft density (threads/cm)	12	14	16	18	20	22
Rmax (cN/thread)	17	24	38	54	68	75
Standard deviation	2	2	3	2	2	2
95% confidence interval	2	1	2	1	1	1

Conclusion

The block diagram below shows maximum values of weaving resistance including an confidence interval.

The figures shows that in the region of lower weft densities the maximum values of weaving resistance of PES and VS materials do not differ a great deal. Larger differences were observed in the region of higher weft densities.

Therefore it can be assumed that the setting of the required weft density based on pure geometric parameters (see Introduction of this article) is reliable in the region of lower and medium weft densities. With higher weft densities, maximum weaving resistance depends on the material used, and hence variations in the achieved weft density for different materials can be assumed. In the region of extremely high weft densities, it is probably advisable to set them on the basis of force quantities in the beat up process.

Figure 7. Tensile forces in warp and in fabric as a function of time, material: PES, warp density 18 threads/cm

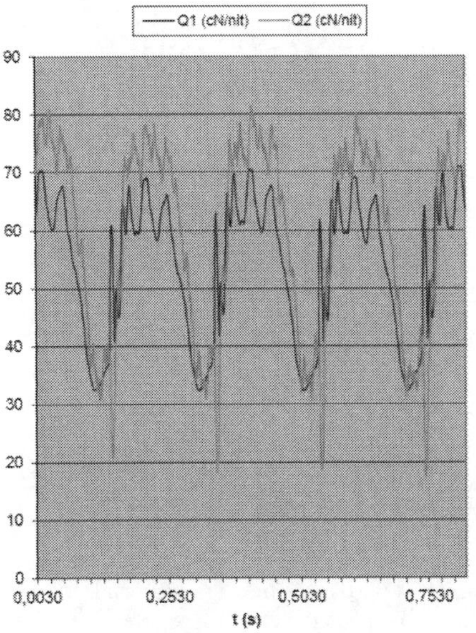

Figure 8. Weaving resistance as a function of time, material: PES, warp density: 18 threads/cm

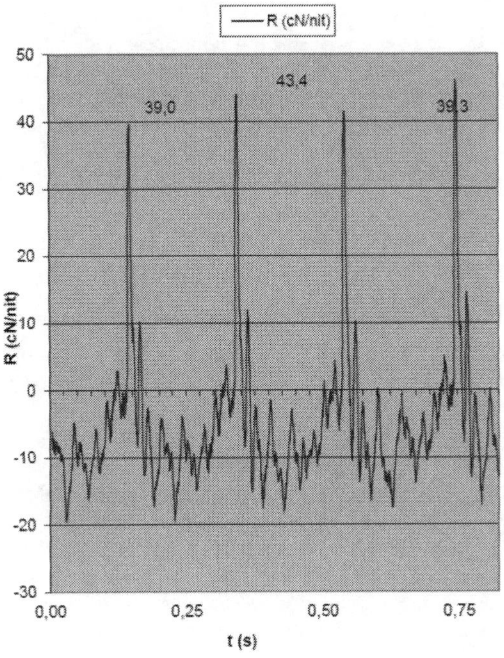

Figure 9: Maximum values of weaving resistance

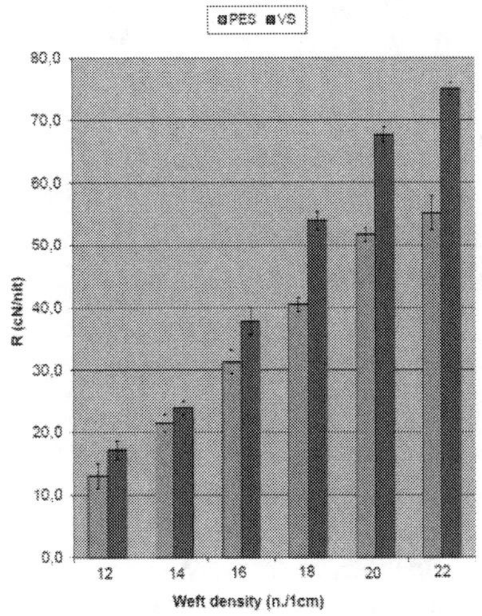

Acknowledgement: This publication was written at the Technical University of Liberec, Faculty of Mechanical Engineering with the support of the Institutional Endowment for the Long Term Conceptual Development of Research Institutes, as provided by the Ministry of Education, Youth and Sports of the Czech Republic in the year 2015.

References

1. Klouček, P., Škop, P.: Použití měřícího zařízení waweon k optimálnímu seřízení textilních, zejména tkacích strojů, 25th International conference quo vadis textrinum, pages: 50-59, 27.4-29.4.2004, NMC spol. s.r.o, Žilina, Slovakia, 2004
2. Morton, W. E. ; Hearle, J. W. S.: Physical properties of textile fibres, Manchester & London, The Textile Institute Butterworths 1962.
3. Technical University of Liberec. Zařízení k měření tkacího odporu na tkacím stroji. Tumajer, P., Bílek, M. Int D 03 D 49/04. Čzech republic. Patent 303018
4. Tumajer P., Ursíny P., Bílek M., Moučková E.; Research Methods for the Dynamic Properties of Textiles. FIBRES & TEXTILES in Eastern Europe 2011, Vol. 19, No. 5 (88) pp. 33-39.
5. Dvorak, J.; Bílek, M; Tumajer, P.: Vývojové trendy tkacích stroju, available from http://tkani.tul.cz/ , ACCESSED 2015-03-09.
6. Nosek, S. At All: Solution of dynamics of weaving process and building the laboratory of diagnostics and measurements of textile machines. International journal of clothing science and technology, 10 (6), PP. 49-51.1998.
7. Tumajer, P., Ursíny, P., Bílek, M., Mouckova, E., Pokorna M.: Influence of structure of the yarn on mechanical characteristics of yarns exposed to dynamic stress, Autex Research Journal, Volume 12, Issue 2, June 2012, Pages 44-49.
8. Pustka M., Kloucek P., Škop P.: Waweon – Instrument for dynamic measurement and analysis of threadand warp tension forces. Melliand International, March 2008. PP. 40-41.

Testing equipment for experimental analysis of mechanical properties of textile materials

Martin Bílek[1a], Šimon Kovář[1], Eva Moučková[2] and Petr Tumajer[2]

[1]Technical University of Liberec, Dept. of Textile Machine Design, Studentská 2, 461 17 Liberec 1, Czech Rep.
[2]Technical University of Liberec, Department of Textile Technologies, Studentská 2, 461 17 Liberec 1, Czech Rep.

Abstract. The regular equipment for testing of textile materials does not facilitate the elongation of threads at a frequency corresponding to the frequency of loading during technological process, and it does not allow for clamping a thread of the corresponding length. In the Technical University Liberec there has been built a testing equipment for experimental analysis of deformation characteristics of textile materials during cyclical loading at a wide range of frequencies (VibTex), and there has been devised the methodology for realisation of experimental measuring and processing of the measured data. The submitted paper explains this experimental equipment for simulation of thread loading, which has been created in the Technical University Liberec for analysis of behaviour of textile materials at high loading frequencies.

1 Introduction

The textile materials and products made of these materials are exposed to varied load regimes. The load regime of finished products made of textile materials is determined by the manner of their use; meanwhile the load regime of textile materials during their processing is determined by technological process. The deformation properties of finished textile products often determine and limit the use of these products in certain areas [1]. The deformation properties of textile materials influence the technological process itself, as well as the quality of the finished product. A typical example is technological process of weaving. In the second half of past century, the output of weaving machines was increasing considerably. Around the turn of the century, this trend has changed, i.e. the outputs of weaving machines are stagnating or decreasing. One of possible reasons of this standstill in the increase of outputs can be deformation properties and behaviour of textile materials in the weaving process. During weaving, the textile material (thread, fabric) is loaded cyclically, at a relatively high frequency. Some products made of textile materials are exposed to cyclical or dynamic load in their use, too, e.g. safety belts, straps, ropes, liquid and air filters, artificial tendons etc.
The dynamic methods of determining deformation and damping properties of the material utilise cyclical deformations of the sinus course primarily. The regular equipment for testing of textile materials (dynamometers) does not facilitate the elongation of threads at a frequency corresponding to the frequency of loading during technological process, and it does not allow for clamping a thread of the corresponding length. In the Technical University Liberec there has been built a testing equipment VibTex for experimental analysis of deformation characteristics of textile materials during cyclical loading at a wide range of frequencies, and there has been devised the methodology for realisation of experimental measuring and processing of the measured data. The constructed equipment allows for ascertaining deformation characteristics of both linear and flat textile objects. It allows also evaluating the energetic intensity when processing varied textile materials (for example on a weaving machine, it is possible evaluate the individual materials, as concerns the energy required for their deformation when opening the shed or during the beat-up process) [3].

2 VibTex: special equipment for testing of textile objects

The equipment VibTex facilitates the simulation of behaviour of textile materials and products exposed to cyclical load. The design of the equipment allows realising the tests of linear and flat textiles at a wide range of frequencies (fig. 2).
The equipment is based on an electromagnetic vibration system, consisting of the proper vibration exciter TIRAvib TV50350, an output amplifier and a double-channelled control unit VR8500. On the vibration exciter, there is fastened an acceleration sensor, and its signal is brought to the first channel of the control unit. By means

[a] Corresponding author: martin.bilek@tul.cz

of the amplifier, the control unit regulates the input voltage to the vibration exciter, in order to achieve the required course of acceleration. The required course of acceleration can be entered by means of the software VibrationVIEW, supplied with the control unit. This software is installed in a regular computer, and it communicates with the control unit by means of network cable.

Figure 1 Diagram of standard connection of the vibration system

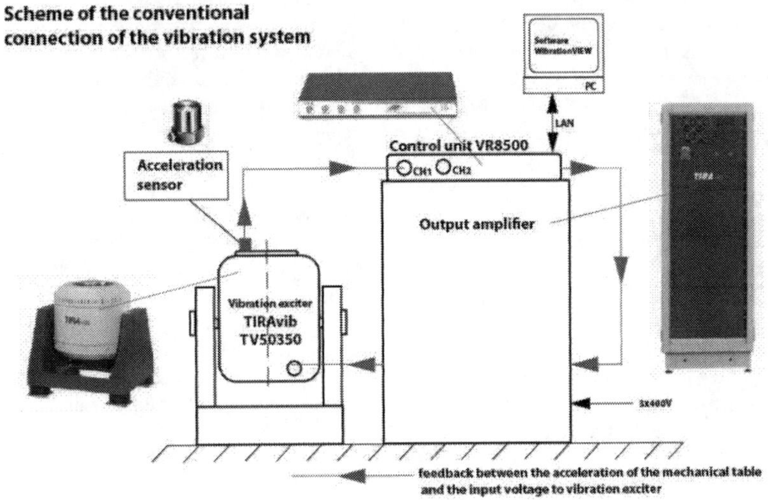

The vibration equipment in the above configuration serves primarily for testing of varied machine components that can be fastened on „mechanical table" of the vibration exciter (fig. 1). For the purpose of testing textile materials, the system must be adapted conveniently. In our case, the vibration exciter is employed as a control member extending the textile sample. Therefore, its axis is oriented horizontally, and upon the „mechanical table" there is fastened the end of the tested sample and an inductance sensor of the path, measuring the elongation of the sample (the exciting function) during the test. The other end of tested sample is fastened to a fixed frame, the construction of which allows adjusting varied clamping lengths of the textile object (from 20 cm up to 3 m). In the fixed frame there is integrated a tensiometer, which allows measuring the tensile force in the textile (the response) during the test. Furthermore, there is situated here the screw for adjusting the required value of „preloading" in the textile sample. Clamping elements for testing both linear and flat textiles are available, too.

Figure 2. Vibration equipment VibTEX

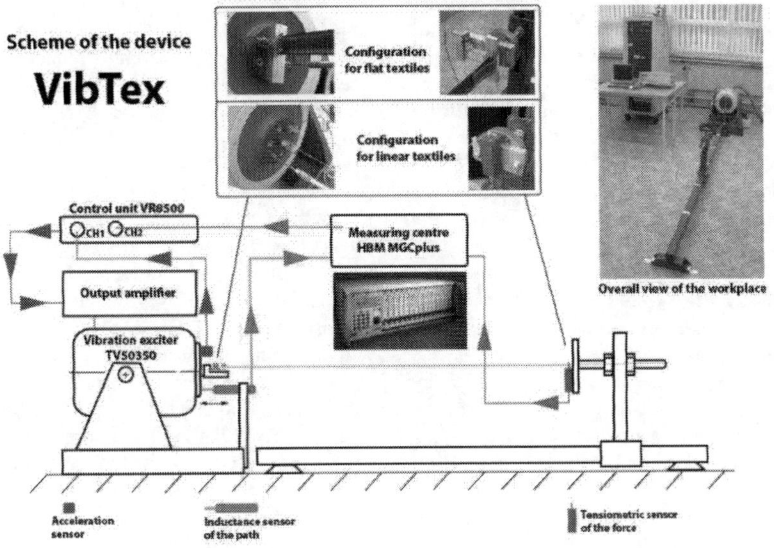

DOI: 10.1051/matecconf/20178901004

The signal coming from both the inductance and tensiometric sensors is processed by means of the measuring centre HBM MGCplus. The signal is digitized here, and the measured values are stored in an ASCII file in respective physical units. Consequently, the main output of measuring is the time dependence of the elongation (exciting function) and the tensile force (response) in the tested sample. The output connector of the board of inductance sensor is interconnected with the other channel of the control unit; therefore, the course of the deflection of „mechanical table" (elongation of the sample) can also be visualised in the program VibrationVIEW.

The software VibrationVIEW allows realising tests with varied courses of elongation. For the formation of tests, two modules are available in the existing configuration: System Check and Field Data Replicator. During routine operation of the vibration system, the module System Check serves for checking its functionality. However, when testing textile objects, this module can be employed for harmonic elongation of the samples successfully. The user interface of this module allows for entering of the required frequency and amplitude of acceleration. At the start of the equipment VibTex, the deviation of the mechanical table goes on increasing gradually up to the required value.

The module Field Data Replicator allows for producing tests from files in varied formats. In case of the testing equipment VibTex, we make use of text files containing two columns of numerals. The first column incorporates the values determining time, and the second one contains the values of required acceleration. The text file can be generated by means of an arbitrary program (e.g. the table calculator Excel), and then exported into the module Field Data Replicator. In this way, it is possible to generate a test with any arbitrary periodical elongation of the textile sample.

The maximum achievable values of elongation (deflections of „mechanical table") are determined by the construction of the vibration exciter. The following figure No. 3 shows the dependence of the maximum achievable deflection on the frequency (in logarithmic co-ordinates). By means of the given vibration system, the textile sample can be elongated by the value 25mm up to 10Hz. After that, the maximum achievable deflection decreases, and at 100Hz, the sample can be elongated by 4 mm approximately.

Figure 3. Maximum deflection

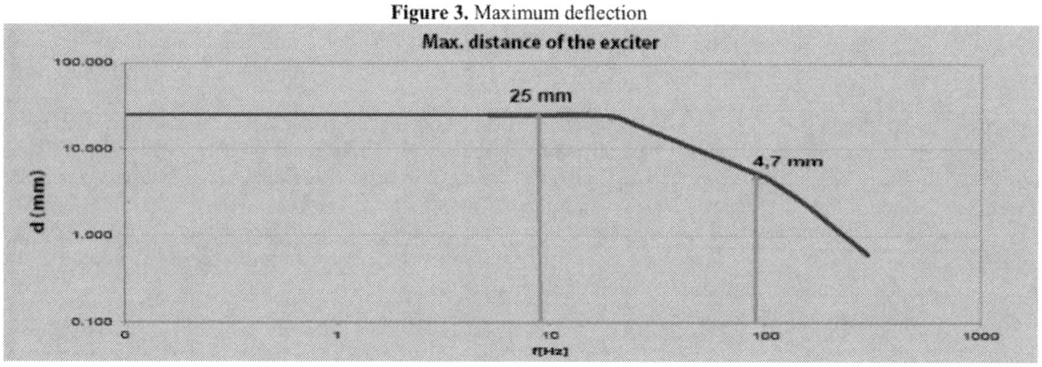

Concerning the evaluation of mechanical properties [2] of tested samples, it is necessary to record the courses of elongation (exciting function) and tensile force (response). As mentioned already above, the signals from the inductance sensor of the path and of the tensiometric sensor of the force are processed by means of the measuring centre HBM MGCplus. By means of a network cable, this centre is interconnected with the computer comprehending the service software HBM Setup Assistant. This software facilitates visualisation of measured data and their storage in text files intended for subsequent further processing by means of the evaluating software VibTex Soft.

In practice, it has come out that the employment of the inductance sensor of the path allows using the testing equipment VibTex up to the maximum frequency 40Hz of the load exerted upon the sample, with the elongation of the sample of textile material 10 mm. Once this operating frequency is exceeded, the record of the path shows instabilities due primarily to the construction of the fixture devised for clamping the inductance sensor of the path to the frame of the vibration exciter. For measuring higher loading frequencies, the inductance sensor of the path can be substituted e.g. by a contactless laser velocity scanner Brüel & Kjær 8338. This laser vibration meter serves for contactless and very fast measuring of vibration velocities. The employed scanner Brüel & Kjær 8338 operates at a distance up to 3 meters from the scanned object.

The basic diagram of measuring has been complemented with a circuit with the record by means of the vibration meter Brüel & Kjær 8338. To the first channel of the measuring centre HBM MGCplus there is brought the signal from the acceleration sensor, and to the second channel, the signal from the inductance sensor of the path. To the third channel, there is brought the signal of velocity from the laser vibration meter. The fig. 4 shows the record of courses of all the three quantities.

The comparison of the courses of the path, velocity and acceleration of harmonic motion confirms a high accordance of the data obtained experimentally and of theoretic courses. At the same time, there is determined

[a] Corresponding author: martin.bilek@tul.cz

 DOI: 10.1051/matecconf/20178901004

the minimum deflection of amplitudes of the deflection, velocity and acceleration in the whole interval of employed exciting frequencies, in comparison with parameters entered by means of the system VibrationView. It is also possible to state a high stability in the adjustment of the exciting frequency. On the basis of this fact, the actual course of elongation of textile material by means of the testing equipment VibTex can be considered ideally harmonic.

Figure 4. Sensing of velocity by means of laser contactless scanner

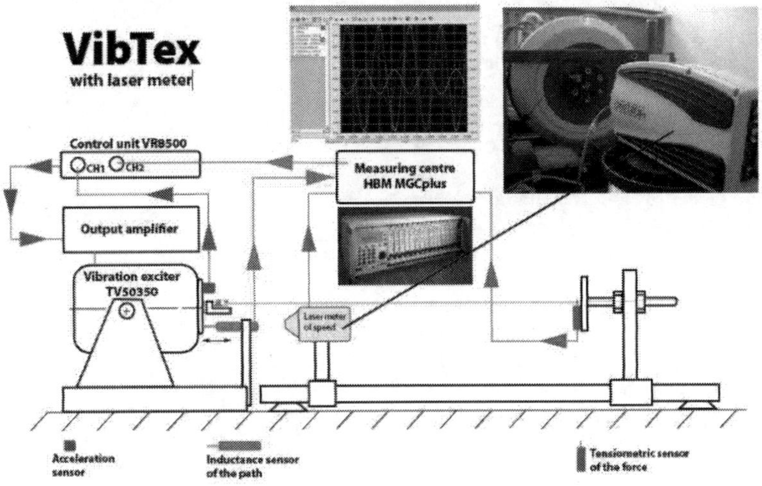

Figure 5. Example of ascertained dependence of the tensile force in the thread upon its elongation at the loading frequency 10Hz with varied values of pre-loading

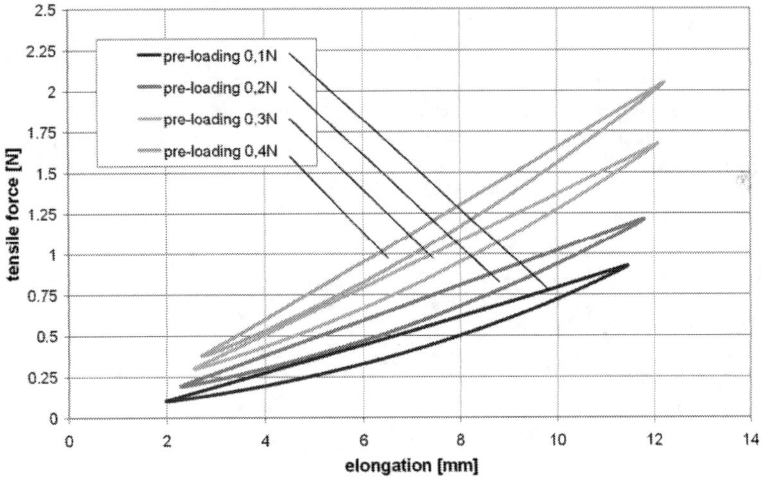

2 VibTex: special equipment for testing of textile objects

The dynamic methods of determining deformation and damping properties of the material utilise cyclical deformations of the sinus course primarily. The body is deformed by the effect of the load. All the work consumed in the deformation of an ideally elastic body is accumulated in it in form of potential energy, which can be recovered on the return to not deformed state. Thanks to viscous properties of material, a part of the work put into the system is converted in heat irreversibly. There exists a number of ways how to express the measure of energy dissipation [5].

If the harmonic motion is steadied, the said dependence can be expressed by a diagram of dependence of the sum of restoring and damping forces of the spring upon the amplitude of the deflection. In case of a visco-elastic substance, this diagram depicts an elliptic shape of the hysteresis loop. An example of obtained dependencies is given in the fig. 5.

Conclusion

As the effected measuring has proven, the testing equipment can be employed for ascertaining the basic mechanical quantities of textile materials with success. It is not problematic to ascertain the forces acting in textile materials. On the contrary, the determination of the actual position of mechanical table of the vibration exciter becomes problematic with this equipment at higher frequencies of loading of the tested sample. This problem can be resolved by employment of a contactless laser vibration scanner.
The testing equipment VibTex allows realising repeatable measuring of deformation properties of textile materials in an ideally harmonic course [4]. On the basic of this measuring, important pieces of information about the examined material can be obtained.

Acknowledgement: This publication was written at the Technical University of Liberec, Faculty of Mechanical Engineering with the support of the Institutional Endowment for the Long Term Conceptual Development of Research Institutes, as provided by the Ministry of Education, Youth and Sports of the Czech Republic in the year 2015.

References

1. Nosek S. Straining of various linear textile bodies in generalized drawing fields, mainly in warps on looms, Conference Textile Science, TU- Liberec, Czech Republic, 1998, pp. 125-138
2. Bílek, M., Brotz, I., Mrázek, J., Tumajer, P.: Mechanical properties of a textile material: Harmonization of mechanic impedances of the weaving machine and the textile material . Buletinul Institutului Politehnic Din Iasi, Mechanica Aplicata 2005. s. 177-184, Iasin, Romania,
3. Bílek, M., Kovář, Š., Mrázek, J., Tumajer, P.: Versucheinrichtung zur Analysierung des Dynamischen Verhaltens des Fadens. XV. Wissenschaftliches symposium TU Liberec – TU Dresden. s. 141-146, Liberec 2005, Czech Republic,
4. Tumajer P., Ursíny P., Bílek M., Moučková E.; Research Methods for the Dynamic Properties of Textiles. FIBRES & TEXTILES in Eastern Europe 2011, Vol. 19, No. 5 (88) pp. 33-39.
5. Morton, W. E.; Hearle, J. W. S.: Physical properties of textile fibres, Manchester & London, The Textile Institute, Butterworths, 1962.
6. Tumajer P., Ursíny P., Bílek M., Moučková E.; Use of the Vibtex vibration system for testing textiles, Autex Research Journal 2011, (11), No 2, pp. 47-53, ISSN 1470-9589
7. Snycerski, M.: The pulsator - a generator of cyclic longitudinal impact loads to simulate weaving conditions for warp yarn, Fibres and Textiles in Eastern Europe, 5 (4), pp. 65-67, ISSN 12303666, 1997.
8. Vlasenko, V., Kovtun, S., Arabuli, A., Bereznenko, S.: Application of the longitudinal resonance vibration method for an investigation of a textile's visco-elastic properties, Vlakna a Textil, 14 (2), pp. 11-14. ISSN 13350617, 2007.

DOI: 10.1051/matecconf/20178901005

Methodology for the Experimental Determination of the Powertrain's Inertia Ellipsoid and its Verification

Pavel Brabec[1,a]

[1]Technical University of Liberec, Dept. of Vehicles and Engines, Studentská 2, 461 17 Liberec 1, Czech Rep.

Abstract. This paper summarizes current scientific knowledge and the results of a study focused on the determination of a powertrain's inertia ellipsoid. The work deals with the application of methods for the experimental determination of the inertia matrix and summarizes their basic potential. The work describes a proposed computational algorithm by means of which the inertia ellipsoid can be determined. The experimental section of the work presents the results of measurements for internal combustion engines (powertrains).

1 Introduction

A number of methods have been used to determine the moments of inertia. All of the methods are based on a principle of relationship between the body's moment of inertia and the frequency of the oscillation. The measurement process assumes the oscillation to be undamped and the time of the oscillation is measured. The fundamental methods for determining the body's moment of inertia are based on the principle of the compound pendulum, torsional suspension unit or bifilar suspension apparatus (and/or trifilar or quadrifilar suspension apparatus, as the case may be).

The work was aimed at propose a methodology for measuring on the basis of which the main mass parameters of the internal combustion engine and transmission system (powertrain) could be identified. The measured powertrain's moment of inertia matrices may be further used in practice as one of the input parameters for the simulation and computation. These data can be primarily used for optimizing the powertrain's flexible mount assembly and simulating the vehicle's crash tests.

An effective solution to the above mentioned problem can be the application of accurate input parameters to the simulation and computation both from the view point of safety (crashtests) and comfort (quality powertrain mount assembly in the car body).

Therefore an experimental stand for determining the mass, center-of-gravity position and moments of a powertrain's inertia was created where necessary measurements (**with the greatest degree of possible accuracy**) were later carried out.

Figure 1. The powertrain´s mount assembly and vehicle´s crashtest simulation results crash [7].

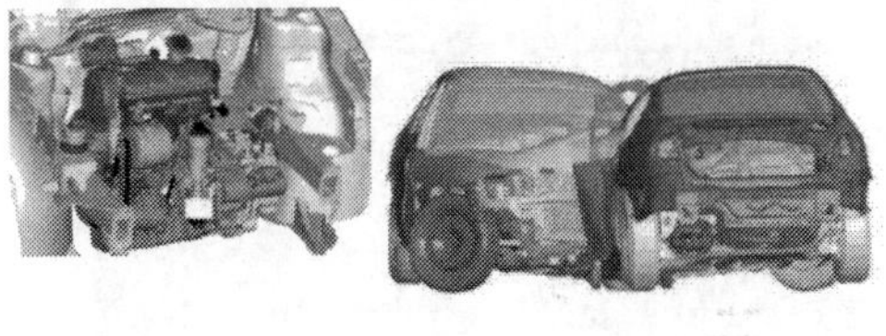

2 The Determination of the Moments of Inertia

The moments of inertia and the deviation moments of inertia can be expressed in a matrix form using one quantity – the inertia matrix,

$$\mathbf{I} = \begin{bmatrix} I_x & -D_{xy} & -D_{xz} \\ -D_{yx} & I_y & -D_{yz} \\ -D_{zx} & -D_{zy} & I_z \end{bmatrix}, \qquad (1)$$

which is the matrix of the second-order symmetrical tensor.

2.1 The Inertia Ellipsoid and Principal Axes of Inertia [2][5]

As the axis passing through the selected origin changes its direction, the body's moment of inertia will change.

[a] Corresponding author: pavel.brabec@tul.cz

Let us plot the reverse square roots of the relevant moments of inertia on these axes from the origin $OM = \sqrt{\frac{1}{I_o}}$ and search for the geometrical locus of end points M. By means of the coordinates (x, y, z) of a point denoted as M, the relationship for the moment of inertia with respect to any axis passing through the origin of the Carthesian coordinate system can be modified to

$$I_x \cdot x^2 + I_y \cdot y^2 + I_z \cdot z^2 - 2 \cdot D_{xy} \cdot x \cdot y - 2 \cdot D_{yz} \cdot y \cdot z - 2 \cdot D_{zx} \cdot z \cdot x = 1 . \qquad (2)$$

The locus sought after is the ellipsoid (generally the triaxial ellipsoid). It is called the **inertia ellipsoid** (hereinafter referred to as ES).

Figure 2. The Inertia Ellipsoid.

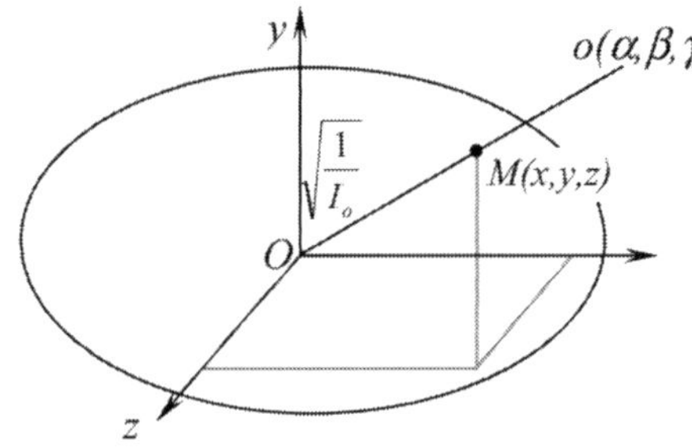

The axes of the ellipsoid are called the principal axes of inertia and then, a condition applies that all three deviation moments with respect to these axes are equal to zero. The moments with respect to these axes are called the principal moments of inertia.

2.2 Experimental Determination of the Moments of Inertia

The moments of inertia can be determined by means of any CAD software if a 3-D model of the component exists. The relevant software requires the bodies'material density values and the appropriate command will compute the centre-of-gravity position and inertia matrices. However, we may face a situation when this approach cannot be used and consequently, the moments of inertia will have to be determined experimentally.

3 Description of Method

Measurement accuracy represented a key aspect for selecting a method for determining the powertrain's inertia ellipsoid. The correctly determined powertrain's inertia matrix is critical for further use in computer simulations, e.g. crashtests.

The determination of the mass and the center of gravity of the powertrain and/or the engine was carried out on the basis of the measurement on a trifilar suspension apparatus by measuring tensile forces in individual wires. To determine the moments of inertia, the indirect method which measures the period of oscillation of the body suspended from a unifilar (torsional) suspension apparatus was used.

To determine the moments of inertia for the bodies having more complex shapes we will definitely need fixtures that will enable suspension of the body in various positions from the torsional suspension apparatus. Therefore, an auxilliary frame for clamping the powertrain (engine) was made. The powertrain was clamped into the auxilliary frame in a defined way so that the axes of the selected powertrain's coordinate system were parallel to the axes of the frame's coordinate system.

Since the frame's moments of inertia are not insignificant, two measurements were needed. The first measurement was made for the frame-powertrain system. The other measurement was carried out separately for the frame so that the frame could be „subtracted" from the measured system. To determine the moments of inertia, the indirect method measuring the period of oscillation of the body suspended from a unifilar (torsional) suspension apparatus was used. The formulae for the computation of the moment of inertia using a torsional suspension apparatus is as follows:

$$I = \frac{G \cdot d^4}{128 \cdot \pi \cdot L} \cdot T^2 \qquad (3)$$

Where I - the moment of inertia, G - spring wire's modulus of elasticity in shear, d - spring wire diameter, L - suspension length, T - time period of one oscillation.

The measurements necessitated development of a suspension apparatus which consisted of universal joints (Cardan joints) and a spring wire. The ends of the suspension apparatus were equipped with carrying yokes into which the spring wire was fixed by clamping. The structure enables the measured body to hang freely (the line passing through the wire axis intersects the body's centre of gravity) for the whole period of the oscillation and for all cases of the suspension (See Fig.3).

The measured quantities were used for the determination of elements of inertia matrix of the powertrain and auxilliary frame. At least six measurements are needed to determine the inertia matrix. However, at least ten measurements were taken to increase the result accuracy based on the relationship determining the moment of inertia with respect to any axis passing through the origin. Consequently, the position of the inertia ellipsoid was determined by means of other mathematical procedures.

If we plot the values of $\sqrt{\frac{1}{J}}$ from the origin using a specific scale as vectors along the axis of the suspension apparatus, the end points will form an ellipsoid of inertia in a space. The following simplified picture shows a process for plotting the inertia ellipsoid in the plane passing through the axis of the crankshaft and axes of the cylinders. Naturally, to enhance the accuracy, more measurements should be taken, ideally a few measurements using the suspension apparatuses which determine the points of the inertia ellipsoid in the proximity of ends of its half-axes (i.e. around the vertices of ellipse).

Figure 3. Principle of plotting an inertia ellipsoid in a plane of the crankshaft.

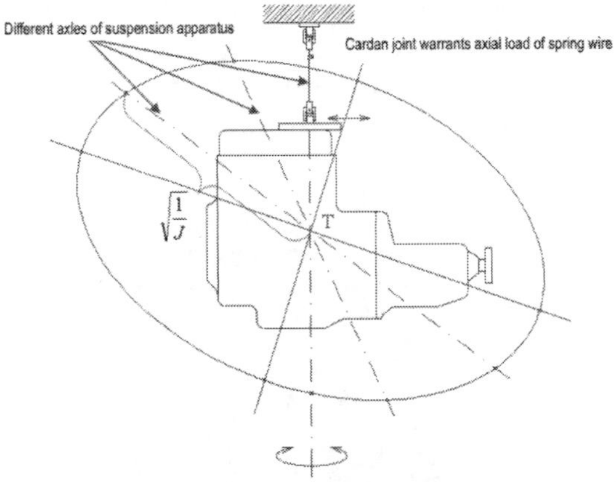

Let us point out that to determine the inertia matrix (i.e. the inertia ellipsoid as well) we have to oscillate the body around at least six different axes using different suspension points. The directions of the individual axes should differ from each other as much as possible. Consequently, we would get six values of the moments of inertia with respect to different axes passing through the centre of gravity and we could determine the inertia matrix of the measured body by means of the a system of equations. Six equations involving six unknowns (using equation (2)) suffice for computing the moments of inertia with respect to the X,Y and Z axes and deviation moments.

For measurement accuracy reasons and for the follow-up computation of the inertia ellipsoid's size and position (hereinafter ZVES – i.e. fundamental properties of the inertia ellipsoid, i.e. the principal moments of inertia and the position of inertia ellipsoid's axes) at least ten measurement of the moment of inertia were taken with respect to any axis passing through the origin. In general, the matrix of n-equations in six unknowns will be obtained for n-measurements. In this event, the unknowns (I_X, I_Y, I_Z, D_{XY}, D_{YZ} and D_{ZX}) can be determined by the method of least squares [8], [9].

The method of least squares is a mathematical method for statistical data processing. It helps find a suitable approximation function for given, empirically determined values. The function being searched for must be a linear combination of the functions which were known in advance and the method will enable their coefficients to be computed. The method of least squares seeks out such values of the coefficients so that the sum of the squares of their functional value deviations from the given measured values was as low as possible. In a more simplified way: „so that the sum of the squares of the deviations was the lowest". We can make up a system of quotations $\mathbf{A}\cdot\mathbf{x}=\mathbf{b}$ ($\mathbf{A}$ is a matrix which we know, $\mathbf{b}$ is a vector which we know, too, $\mathbf{x}$ is an unknown vector).

$$\mathbf{x}=\begin{bmatrix} I_X \\ I_Y \\ I_Z \\ D_{XY} \\ D_{YZ} \\ D_{ZX} \end{bmatrix}, \qquad \mathbf{b}=\begin{bmatrix} I_{o1} \\ I_{o2} \\ \dots \\ \dots \\ I_{on} \end{bmatrix},$$

$$\mathbf{A}=\begin{bmatrix} \cos^2\alpha_1 & \cos^2\beta_1 & \cos^2\gamma_1 & -2\cdot\cos\alpha_1\cdot\cos\beta_1 & -2\cdot\cos\beta_1\cdot\cos\gamma_1 & -2\cdot\cos\gamma_1\cdot\cos\alpha_1 \\ \cos^2\alpha_2 & \cos^2\beta_2 & \cos^2\gamma_2 & -2\cdot\cos\alpha_2\cdot\cos\beta_2 & -2\cdot\cos\beta_2\cdot\cos\gamma_2 & -2\cdot\cos\gamma_2\cdot\cos\alpha_2 \\ \dots & \dots & \dots & \dots & \dots & \dots \\ \dots & \dots & \dots & \dots & \dots & \dots \\ \cos^2\alpha_n & \cos^2\beta_n & \cos^2\gamma_n & -2\cdot\cos\alpha_n\cdot\cos\beta_n & -2\cdot\cos\beta_n\cdot\cos\gamma_n & -2\cdot\cos\gamma_n\cdot\cos\alpha_n \end{bmatrix}$$

In the present case the matrix $\mathbf{A}$ has more rows than columns, the size of the vector $\mathbf{b}$ is bigger than the one of the vector $\mathbf{x}$. It is an over determined system of equations which has more equations than unknowns. In this event it is theoretically possible to ignore some equations but let us suppose that the equations result from experimental data where each measurement is considered valuable and we are not able to decide which of these laboriously obtained equations should be ignored.

In the cases of the over determined system of linear equations such a system has an infinite number of solutions (or more precisely, no solution). Let us say that such a fact is not satisfactory and define an error in solution $\mathbf{e}=\mathbf{A}\cdot\mathbf{x}-\mathbf{b}$.

We want to find such a solution $\mathbf{x}$, so that the error (or more precisely the size of the vector $\mathbf{e}$) was minimal. The solution of our system of equations can be derived by derivation of matrices and vectors. Since $\mathbf{e}$ is a vector, let us adapt the requirement so that the sum of the squares of the individual deviations (i.e. the elements of the vector $\mathbf{e}$) is „minimal". Defining a criterium in such a way actually results in the minimization of the scalar product which can be written by means of transposition.

$$\mathbf{e}^T\cdot\mathbf{e}\rightarrow\min$$

The product will be minimal if its derivation according the variable will equal zero.

$(\mathbf{e}^T\cdot\mathbf{e})'=\left[(\mathbf{A}\cdot\mathbf{x}-\mathbf{b})^T\cdot(\mathbf{A}\cdot\mathbf{x}-\mathbf{b})\right]'=0$. The relationship can be further modified using the rules for transposition of products and derivation of the products of matrices and vectors.

$$\left[\mathbf{x}^T\cdot\mathbf{A}^T\cdot\mathbf{A}\cdot\mathbf{x}-\mathbf{x}^T\cdot\mathbf{A}^T\cdot\mathbf{b}-\mathbf{b}^T\cdot\mathbf{A}\cdot\mathbf{x}+\mathbf{b}^T\cdot\mathbf{b}\right]'=2\cdot\mathbf{A}^T\cdot\mathbf{A}\cdot\mathbf{x}-2\cdot\mathbf{A}^T\cdot\mathbf{b}=0$$

The relationship above may be used to express the resultant formulae for $\mathbf{x}$.

$$\mathbf{x} = \left(\mathbf{A}^T \cdot \mathbf{A}\right)^{-1} \cdot \mathbf{A}^T \cdot \mathbf{b} \qquad (4)$$

Determination of the centre-of-gravity position was made by means of measurement on a trifilar suspension apparatus (measuring the tensile forces in individual wires). Firstly, the centre-of-gravity position was determined in one plane XY, then the frame (or the powertrain-frame assembly) turned and the centre-of-gravity position was determined in the plane perpendicular to the original plane (YZ).

The measurement described above i.e. the determination of the centre-of-gravity position from the origin along all three axes will be carried out separately for the frame as well as for the assembly (the frame and the powertrain). Consequently, the measured and computed values can be used to compute the mass and the position of the powertrain's centre-of-gravity.

The last computational operation which has to be performed to determine the inertia matrix of the powertrain (engine) is subtraction of the auxilliary frame's inertia matrix from the frame-powertrain assembly's inertia matrix. Steiner's theorem will be used for this computation.

Figure 4. Illustration of the measured inertia ellipsoid.

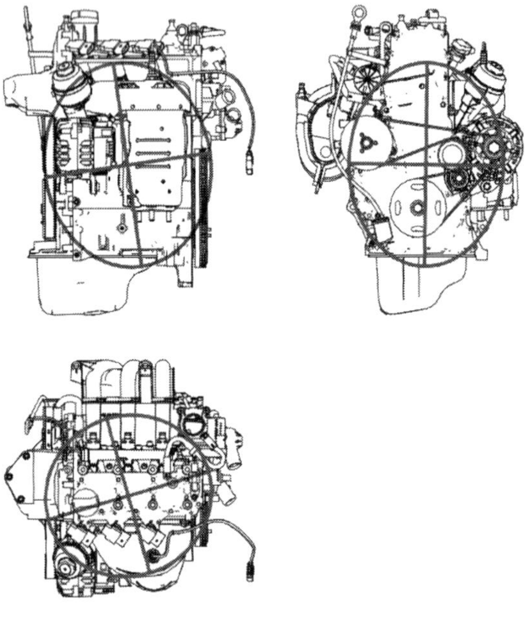

4 Measurement Accuracy

4.1 Measurement verification using a simple body clamped in the frame

To verify the measurement method the following procedure was selected. Firstly, a geometrically simple body was chosen and the body passed through the complete measurement process as a separate powertrain – i.e. firstly, the frame-body assembly and then the frame itself were measured. A simple form of the body was selected due to a simple calculation of the body's moment of inertia matrix which was further compared with the measured matrix. Considering the semi-finished products being available in our laboratory, an assembly consisting of two semi-finished products (Ø100 and Ø130 mm) was chosen. The position of the semi-finished products was chosen in such a way so that one of the deviation moments was not zero. See the Figure below.

Figure 5. Semi-finished products for the measurement verification.

The following evaluation of the results

$$\Delta = \frac{H_N - H_V}{H_V} \cdot 100 \ \ [\%] \qquad (5)$$

(H_N – measured value; H_V – computed value using 3D CAD), will result in the following relative errors in measurements

$$\begin{matrix} I_X & I_{XY} & I_{XZ} \\ I_{YX} & I_Y & I_{YZ} \\ I_{ZX} & I_{ZY} & I_Z \end{matrix} = \begin{vmatrix} 0{,}3321 & & -2{,}5298 \\ & 1{,}1068 & \\ -2{,}5298 & & -0{,}4523 \end{vmatrix} \% \cdot$$

The principal moments of inertia showed very good conformity, the most significant deviation occurred along the x-axis and it was less than **1.5 %**.

The position of the semi-finished products can be considered asymmetrical. However, the asymmetry is chosen in such a way that two deviation moments are zero. For these two deviation moments the relative error cannot be determined, we are able to determine only the absolute one (i.e. $-0{,}0895\, kg \cdot m^2$ and $-0{,}0198\, kg \cdot m^2$). The last deviation moment was measured approximately with the relevant error of **2.5 %**. This result is very good taking into account the comparison of the deviation moments'sizes (comparing with the principal moments of inertia, lower orders of values are in question).

4.2 Experimental determination of the position of the principal axis of inertia for a semi-finished product situated „across“ the frame

By further measurement to verify the method, a semi-finished product having the dimensions of Ø60-1500 mm was located into the position of the auxilliary frame diagonal. The measurement method matched the verification described in the previous section (i.e. the method which included the powertrain as well). Firstly, the moments of inertia matrix of the chosen semi-finished product was computed using the CAD software and then it was determined by means of measurements. These moments of inertia matrices served for further determination and comparison of the solid angles of the principal axis of inertia from the coordinate system. See the following illustration.

Figure 6. Photograph and CAD model with a semi-finished product located diagonally.

Using the same evaluation of the results as for previous cases (see formula 5), we can write: the following relative errors of measurement:

$$\begin{matrix} I_X & I_{XY} & I_{XZ} \\ I_{YX} & I_Y & I_{YZ} \\ I_{ZX} & I_{ZY} & I_Z \end{matrix} = \begin{vmatrix} -0{,}2455 & 0{,}6231 & -0{,}6673 \\ 0{,}6231 & -2{,}5425 & 1{,}4292 \\ -0{,}6673 & 1{,}4292 & 1{,}1100 \end{vmatrix} \%,$$

$$\begin{matrix} \alpha_1 \\ \beta_1 \\ \gamma_1 \end{matrix} = \begin{vmatrix} 0{,}2923 \\ -1{,}2953 \\ 1{,}1467 \end{vmatrix} \%.$$

A comparison of the computed and measured moments of inertia values resulted in an average error of **1.10 %**, the max. deviation occurred along the y-axis (**2.54 %**). The absolute error along the principal axis of inertia (identical with the semi-finished product axis) was less than 1° for all solid angles.

4.3 The influence of the air filter on the inertia ellipsoid of powertrain for passenger car

Further measurements were carried out two variants of the powertrain, which consisted of 1.4 59 kW MPI engine and gearbox MQ 200. First unit was measured without the air filter and later was measured second variant with intake filter. Weight of the plastic intake filter was 1.17 kg (which is approximately 0.9 % of the weight of the aggregate).
Even with such a small change in weight (but at a great distance from the center of gravity) occurred in all ten measuring of the oscillation period of the assembly (i.e. frame + powertrain) in a different position to a measurement a long time for time period of one oscillation of powertrain with intake filter. Differences measured times one oscillation were around 0.6 %. Measured values are shown in the following table.

Table 1. Spatial angles and times of one oscillation.

α (°)	β (°)	γ (°)	without filter T_{BF} (s)	with filter T_{SF} (s)	$[(T_{BF}-T_{SF})/T_{BF}].100$
62,17	53,58	131,17	10,822	10,89	0,63
122,75	55,75	128,08	10,588	10,626	0,36
63,42	130,42	127,83	9,373	9,449	0,81
120,83	128,25	126,67	9,969	9,988	0,19
122,67	130,33	57,67	10,76	10,835	0,70
93,5	49,17	138,92	10,529	10,583	0,51
54,17	142,33	80,08	9,753	9,83	0,79
92	13,25	104,17	9,374	9,447	0,78
5,42	94,08	93,83	11,217	11,267	0,45
94,08	94,42	6,08	10,69	10,725	0,33

If these changes in inertia matrix are expressed as a percentage according to the formula $[(I_{SF} - I_{BF})/ I_{BF}]\cdot 100$, so we come out

$$\begin{vmatrix} 1{,}33 & 1{,}52 & -49{,}79 \\ 1{,}52 & 3{,}84 & 1{,}35 \\ -49{,}79 & 1{,}35 & 0{,}65 \end{vmatrix} [\%].$$

5 Conclusions

The presented work deals with an experimental determination of the body's moment of inertia. The work focuses on evaluating the data obtained from measurements carried out on various types of powertrains or engines. To meet the objectives, a measurement method to determine the moments of inertia, centre-of-gravity position and the powertrain mass was developed. Then specialized aids for measurements were developed together with a special program for evaluating the measured data.

The method was verified by measuring several powertrains of engines, in some cases for more versions of the engine accessory arrangement (twelve powertrains and one engine used for passenger cars, one engine used for a commercial vehicle). The measurement was always performed without media i.e. without oil and coolant.

The metering of inertia moment is well known, but measurement accuracy represented a key requirement in the measurement method design. Consequently, more values were always measured and for the determination of the ellipsoid of inertia an approximation using the method of least squares was used.

The mathematical derivation which considered all adverse impacts to a maximum extent resulted in approximately 6 % measurement accuracy. A simple-shaped body was used for the complete measurement process calibration. The relative error between the measured values and the CAD program computed values was up to 2.5 %.

The mass (in the form of engine's accessories) being added to or removed from the engine affects the resulting values of the moments of inertia and positions of the centres of gravity. The change in the moment of inertia depends on the following factors: the moment of inertia of the relevant component, the mass (m) and the distance from the centre of gravity (e). The relation of $I_0 = I_{0S} + m \cdot e^2$ is known as the Steiner formula [2]. The distance from the centre of gravity plays a significant role because even a relatively small mass can affect the total moment of inertia of the power train. Several versions of measurements of engines and powertrains with varied arrangement were chosen to assess the impact of accessories. E.g. the measurement carried out on the truck engine (3.8 l Avalon Engine) resulted in nearly 20 % change in the axial moment of inertia when approximately a 10 % change in mass occurred.

Figure 7. Comparison of the inertia ellipsoid of the engine with and without a compressor for air conditioning.

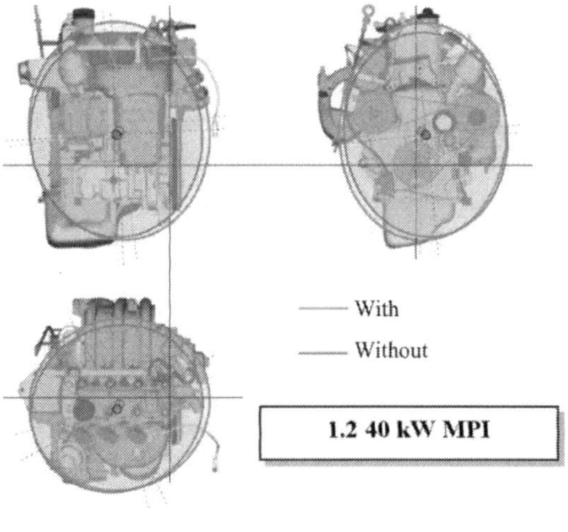

The proposed measurement methodology was able to reveal even a relatively small difference in measurement depending on the powertrain's accessories. Consequently, the change resulting from removal of the engine's intake filter (1.4, 59 kW MPI engine with MQ 200 gearbox) which formed 0.9 % of the total powertrain's mass, could be observed.

Further work could focus on more detailed analysis of certain issues, impact of engine's liquid media, simplification of the method (its laboriousness) while maintaining the measurement accuracy.

References

1. F. Vlk, Zkoušení a diagnostika motorových vozidel, Nakladatelství a vydavatelství vlk, BRNO (2001)
2. Z. Bradský, R. Vrzala, Mechanika III. /Dynamika/, VŠST, Liberec (1990)
3. 15. Aachen Colloquium, EUROGRESS Aachen (2006)
4. 5. Tag des Fahrwerks, Institut für Kraftfahrwesen Aachen – RWTH Aachen (09.10.2006)
5. Technická zpráva firmy Škoda Auto At 3374-D, odd. 1-231
6. Podklady firmy Škoda Auto
7. www.auto.cz
8. M. Brzenina, M. Dvořák, Z. Kalousek, P. Salač, J. Staněk, Matematika 4, TU v Liberci, Liberec (1996)
9. www.wikipedie.cz
10. H. Gortz, Identifikation von Fahrzeugträgheitsparametern in Fahrversuchen und auf Prüfständen, Institut für Kraftfahrwesen Aachen der RWTH Aachen, ISBN 978-3-925194-89-4
11. A. Best, Moment of inertia measurement system, MIMS Description and Specification, Anthony Best Dynamics (2004)
12. R. Sharp, The Measurement of Mass and Inertial Properties of Vehicles and Components, Cranfield University, Cranfield UK (1997)
13. D. Andreatta, G. Heydinger, R. Bixel, D. Coovert, Inertia measurements of large military vehicles, 2001 SAE World Congress Detroit (2001-01-0792)
14. P. Brabec, R. Voženílek, Elipsoid setrvačnosti agregátu - Škoda Auto 1.6 76kW TDI (DPF) Common Rail + MQ250. ZPRÁVA SM595/2008. TU v Liberci, Liberec (2008)
15. Pavel Brabec, Experimentální určení polohy a rozměrů elipsoidu setrvačnosti hnacího agregátu. Teze disertační práce. Technická univerzita v Liberci. (2009)

Effects of injection and ignition timing of a large direct injection spark ignition engine fuelled with a hydrogen-CNG mixture on its operating conditions and emissions

Michael Fenkl[1,a]

[1]Technical University of Liberec, Dept. of Power Eng. Equipment, Studentská 2, 461 17 Liberec 1, Czech Rep.

Abstract. The effects of ignition and fuel injection timing of an experimental 2-liter single cylinder direct injection spark ignition gaseous fuel engine were examined experimentally in this study. An online-prepared mixture of 10-40% hydrogen with compressed natural gas was injected at 8 bar into the cylinder during the intake stroke at 1500 rpm. Heat release rates, maximum indicated pressures, cycle-to-cycle variability, and gaseous emissions were compared for different settings.

1 Introduction

The efforts to reduce the effects of exhaust emissions from internal combustion engines on environment and human health, and the resulting drive towards environmentally friendly power generation and transportation, motivate research groups and engine manufacturers to seek suitable alternative fuels, which can be produced from renewable resources, and which fulfill the increasingly demanding requirements for exhaust emissions. Such effort is not limited to vehicle engine, but also applies to stacionary engines used in distributed power generation in combined heat and power plants.

Compressed natural gas (CNG), composed of primarily methane, is a popular and widely used alternative fuel, which can be combusted with very low exhaust emissions compared to gasoline or diesel fuel [1]. Unlike automobile engines, which run at stoichiometric air-fuel ratio and use a three-way catalyst, larger power generation engines run lean, at an air-fuel ratio where the engine-out emissions of nitrogen oxides (NO_x) are relatively low, while the combustion efficiency is still acceptable. Such approach allows for the use of biogas, which often contains impurities which are expensive to remove and which would poison a catalyst. Addition of even small percentage of hydrogen into the fuel was reported to substantially improve the fuel properties [1-5], especially the lean combustion limit, improving also the initial stages of combustion, NO_x emissions and/or combustion efficiency.

Compared to gasoline and diesel fuel, methane already offers lower CO2 emissions. These are further decreased by displacement of methane with hydrogen. Extension of the lean combustion limit due to hydrogen addition allows the engine to run leaner while maintaining comparable combustion quality [1], allowing to reduce combustion temperatures which have a strong effect on NOx. A possible downside of hydrogen addition is lower volumetric energy density of hydrogen relative to methane and liquid fuels, resulting in displacement of combustion air with fuel.

Ma [3] reports that on a naturally aspirated 6.2-liter, six-cylinder engine, addition of 20% hydrogen by volume improved both the mean indicated pressure variability and the maximum indicated pressure, while the emissions of hydrocarbons (HC) were decreased but NOx slightly increased. Huang [4] reports that addition of hydrogen to CNG allows for a significant increase in exhaust gas recirculation (EGR) rate while maintaining low indicated pressure variability. Thurnheer [5] reports that addition of 15% hydrogen by volume has allowed, compared to methane only, to delay ignition timing by 4.5 crank angle degrees (CAD).

2 Experimental setup

2.1. Engine and instrumentation

The experiments were carried on a unique single-cylinder research engine, made from a 12-liter, six-cylinder LIAZ ML 636 diesel engine. The cylinder no. 6 was equipped with a modified piston, lowering the compression ratio to 12:1. The cylinder head was fitted with a fuel injector (used in place of a glow plug), with a spark plug (used in place of a diesel injector) with an ignition module, and with a cylinder pressure sensor. The remaining five cylinders were used to pump air into a reservoir feeding the working cylinder, simulating

[a] Corresponding author: michael.fenkl@tul.cz

DOI: 10.1051/matecconf/20178901006

turbocharging. The parameters of the engine are given in Table 1. All engine modifications were done in-house and were described in [6-8]. Hardware and software for controlling the fuel injection and ignition was fabricated in-house. The schematics and photo of the engine is given in Fig. 1. The photo of the fuel injector is given in Fig. 2.

The engine was coupled to a DC electric dynamometer, allowing for both motoring and braking of the engine. The engine was operated at 1500 rpm, a nominal speed of engines used in medium-size combined heat and power plants. The engine was not equipped with throttle; the power output was regulated by changing the air-fuel ratio.

Figure 1. Engine and test stand schematics.

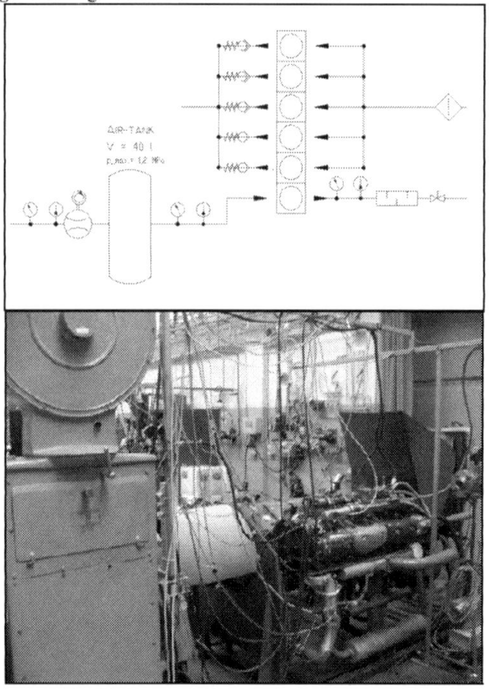

Figure 2: Fuel injector

Table 1. Engine specifications.

Type of engine	naturally aspirated SI engine
Number of cylinders (-)	1
Displacement volume (L)	1,99
Compression ratio (-)	12
Bore (mm)	130
Stroke (mm)	150
Connecting rod length (mm)	250
Intake valve timing	IVO: 3° BTDC IVC: 44° ABDC
Exhaust valve timing	EVO: 53° BBDC EVC: 7° ATDC

The in-cylinder pressure was measured by a piezoelectric indicated pressure sensor (GU 21D, AVL), connected to a base unit (Indimetr 619, AVL), and analyzed by Concerto software (AVL). Other engine parameters were measured at 1-second intervals with a data acquisition system (cDAQ-9172, with modules NI 9217, NI 9211, NI 9202, NI 9205, National Instruments) and processed by an in-house written software using the LabView application (National Instruments).

The methane content of the fuel was measured by an NDIR analyzer, developed, calibrated and validated in earlier work by this group [9]. The concentrations of HC, CO, NOx and CO2 in the exhaust were measured by an in-house developed portable emissions monitoring system [Vojtíšek SAE 2009-24-0148].

2.2. Fuel mixture preparation

Hydrogen and CNG fuels were taken from the high-pressure laboratory fuel supply system. Natural gas was taken from the distribution network, compressed by a fueling station compressor and stored in CNG tanks. Bottled compressed hydrogen was purchased (Linde Gas) and stored on site. Both gases were passed through Coriolis-type fuel consumption meters (Micro Motion Elite CMF 010) a pressure regulators set to 2.5 MPa feeding an in-house developed mixing device, allowing online mixing of the fuels with on-the-fly adjustment of the mixture composition. The device is desrcibed in Czech patent no. 298642 [10], and shown in Fig. 3. The resulting fuel mixture was analyzed using an in-house NDIR analyzer [9]. The fuel supply system schematics is presented in Fig. 4.

DOI: 10.1051/matecconf/20178901006

Figure 3. Fuel mixing device.

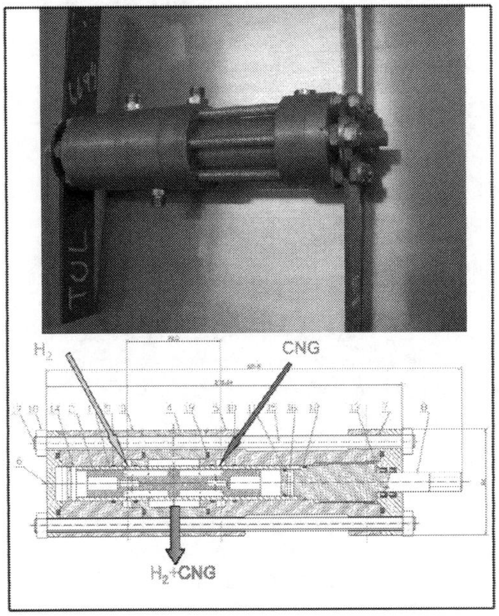

Figure 4. Fuel supply system.

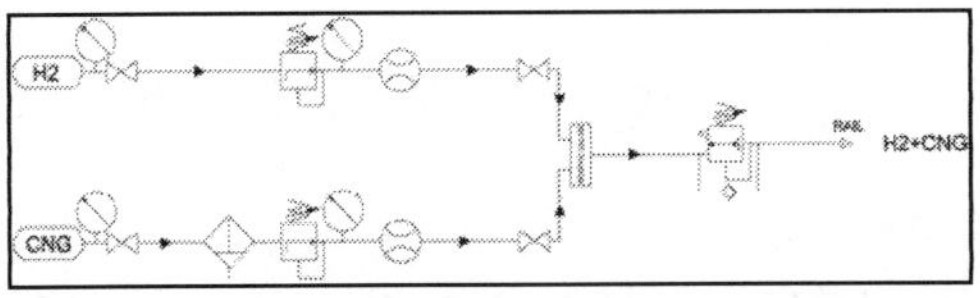

3 Results and discussion

Throughout this section, unless indicated otherwise, the fuel injection timing took place during the intake stroke and is measured in degrees after the top dead center (ATDC), while the ignition timing (during the compression stroke) is measured in degrees before the top dead center (BTDC).

3.1 Ignition timing effects

Out of the relatively large set of aggregate data, operating points with comparable excess air ratio of 1.33, comparable fuel composition of 13-14% H2, and comparable injection timing at 50 ATDC (40 ATDC for ignition timing 24 BTDC).

It is apparent from Fig. 5 that as the ignition timing is advanced, the concentrations of NOx increase, while HC and CO decrease, and CO2 remains constant. (The small deviation in the fuel composition can be considered insigificant according to [3].) These results are in agreement with another study [3]. The relative changes to 22 BTDC are given in Fig. 6.

Figure 5. Effects of ignition timing on exhaust gas composition.

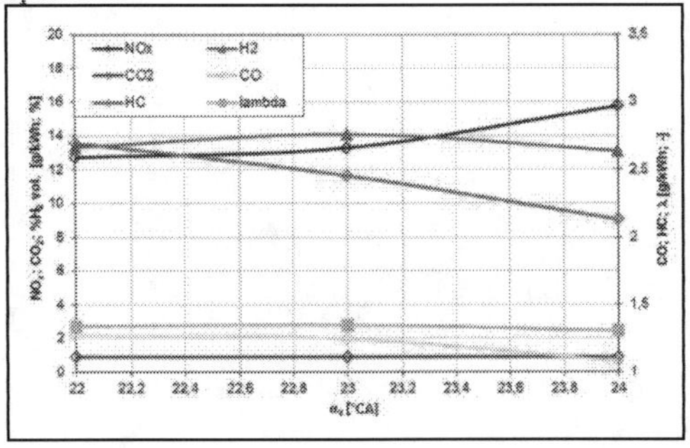

Figure 6. Relative changes in concentrations of NOx, CO2, CO, HC as a function of ignition timing.

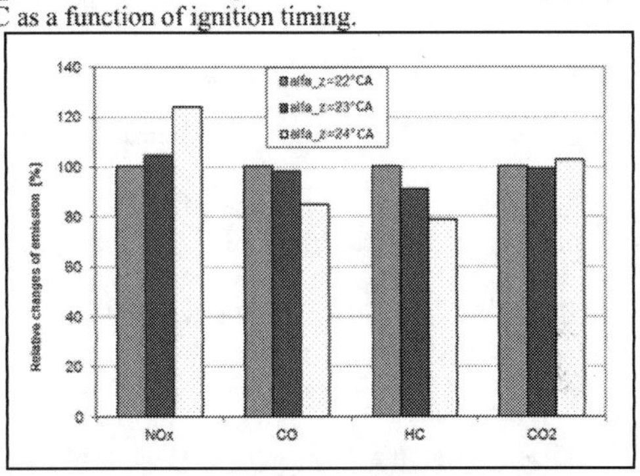

Also, as the ignition timing is advanced, the duration of the main phase of the combustion (defined as the period from 10% to 90% of heat released) is longer, as shown in Fig. 7. This allows for improved combustion efficiency and decrease in HC and CO emissions. Detailed examination of the heat release progress, shown in Fig. 8, suggests that this change is attributed nearly entirely to the second phase of combustion (50-90% heat released), while the first phase of combustion (10-50% heat released), and also the sum of ignition delay and the initial phase of combustion (from nominal ignition angle to 10% heat released) are comparable across all ignition timings examined.

Figure 7. Crank angle at which 10%; 50% and 90% of the energy in the fuel is released, and the length of the intervals 10-50%, 50-90% and 10-90%, as a function of ignition timing.

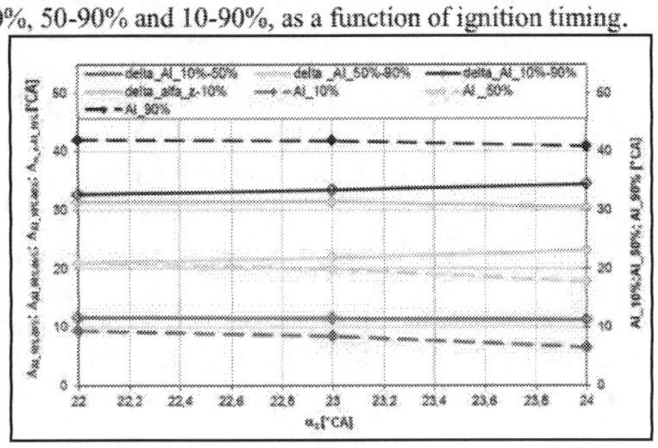

Figure 8. Heat release rates as a function of ignition timing.

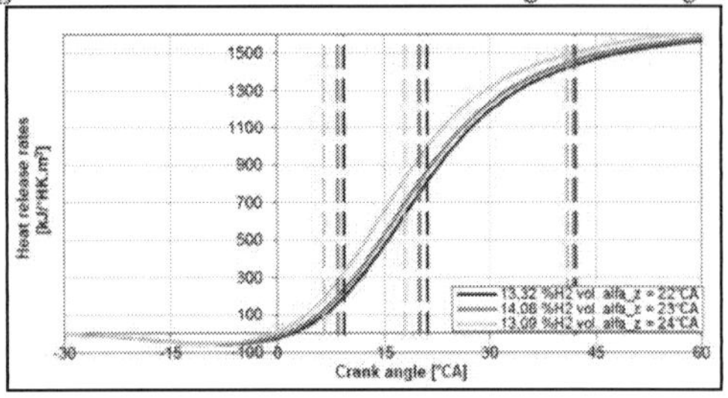

From these data it can be ascertained that the onset of the oxidation reactions is not majorly influenced by the ignition timing and that the ignition timing influences primarily the second phase of the combustion.

3.2 Effects of start of the injection timing

Three groups of data, each of which composed of points with comparable air-fuel ratio, hydrogen content of fuel, and ignition timing, have been examined.

In the first group, fuel containing 31% of hydrogen by volume was injected at 35-40 ATDC, the ignition timing was 27 BTDC, and the excess air ratio was 1.68.

The second group spans injection timing 50-80 ATDC, excess air ratio of 1.58, 34.6% hydrogen by volume, and ignition at 27 BTDC.

In the third group, the fuel was injected at 100-120 ATDC, contained 30.7% hydrogen by volume, excess air ratio was 1.48, and ignition timing 27 BTDC.

Figure 9. Concentrations of NO_x, CO_2, CO, HC as a function of injection timing.

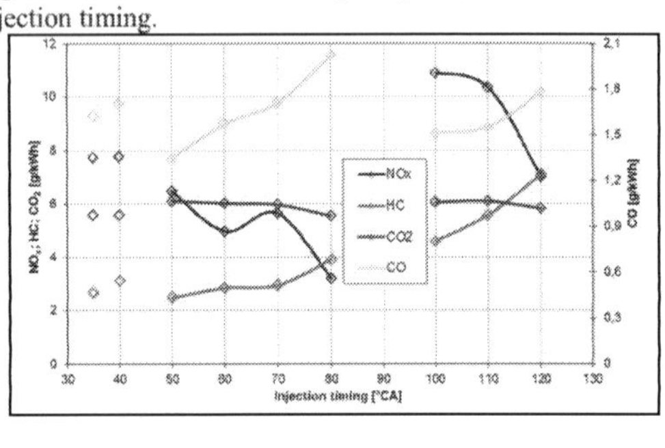

For the second group, data in Fig. 9 suggest that with advancement of the start of the fuel injection, HC and CO decrease considerably, NOx increases, and there is very little increase in CO2. This is in agreement with [11], in which it has also been found that around 120 ATDC the NOx trend reverses and NOx moderately increase with further delayment of the start of the injection up to the bottom dead center, after which NOx dramatically decreases.

The step change from 80 to 100 ATDC is attributed to differences in air-fuel ratio and hydrogen concentration, as shown in Fig. 10.

Figure 10. Effect of start of the injection on maximum and mean indicated pressure.

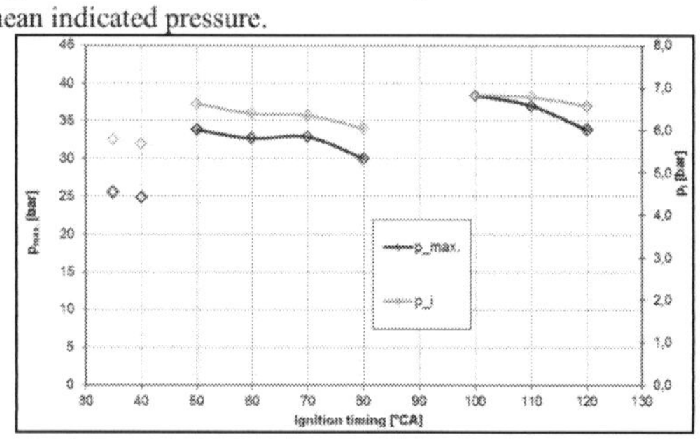

Later injection decreases the time available for mixing of the fuel with the air, reducing the mixture homogenity, with a resulting increase in HC and CO. This is in agreement with the observed increse in variability of the mean indicated pressure, plotted in Fig. 11. This claim is also supported by the shift of the combustion onset, which, with delayment of the injection, is shifting away from the ignition (increased ignition delay), as apparent in Fig. 12. Delaying the start of injection is therefore believed to result in the degradation of the homogenity of the mixture. While such change may be insignificant compared to comparable effects on liquid fuels, it is well observable, as demonstrated in measurable changes in parameters indicating onset and quality of the combustion. The effect on mixture homogenity is also discussed in [11].

Figure 11. Effect of start of the injection on variability in maximum and mean indicated pressures.

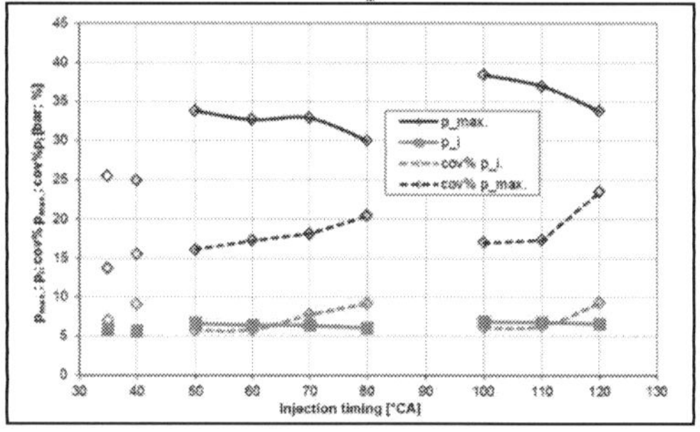

Figure 12. Crank angle at which 10%; 50% and 90% of the energy in the fuel is released, and the length of the intervals 10-50%, 50-90% and 10-90%, as a function of timing of the start of the fuel injection.

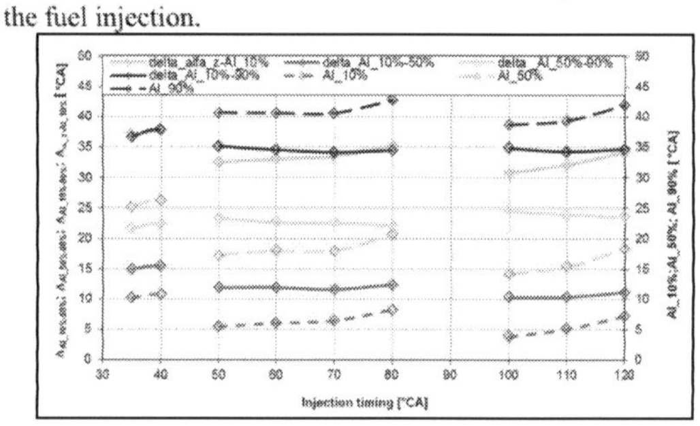

4 Conclusions

- With the advancing ignition timing, while keeping the air-fuel ratio and fuel injection timing constant, the emissions of NOx increased, HC and CO decreased, and CO2 did not change significantly. There was no detectable change in the onset of combustion, its early stage, and its main stage until 50% heat released; time from 50% to 90% heat released increased with advanced ignition timing. The comparable timing of the first half of the combustion suggests that the ignition timing has little effect on pre-combustion and early combustion reactions. The longer duration of the second half of the combustion is believed to have resulted in more complete combustion and observed reductions in HC and CO emissions.
- Delayment of the start of the fuel injection, while ignition timing and air-fuel ratio are kept constant, has resulted in lower NOx and higher HC and CO emissions, while CO2 concentrations were comparable.
- Characteristic points of heat release rate are significantly delayed with increasing injection delay in each area with the constant ignition timing; air-fuel ratio and H2/CNG were significant. The initialization time (°CA) between ignition timing and Al_10% is significant depended on start of injection while the time between Al_10% and Al_50% is approximately constant and the time between Al_50% and Al_90% decreases.

References

1. G.A. Karim, I. Wierzba, Y. AlAlousi, Methane–hydrogenmixtures as fuels, J Hydrogen Energy . (1996)
2. M. Fanhua, W. Yu, Study on the extension of lean operation limit through hydrogen enrichment in a natural gas spark-ignition engine, International journal of hydrogen energy, **vol. 33**, p. 1416–1424, (2008)
3. M. Fanhua, W. Yu, L. Haiquan, L. Yong, W. Junjun, D. Shangfen, Effects of hydrogen addition on cycle-by-cycle variations in a lean burn natural gas spark-ignition engine, International journal of hydrogen energy **vol. 33**, p. 823–831, (2008)
4. H. Erjiang, H. Zuohua, L. Bing, Z. Jiajun, G. Xiaolei, Experimental study on combustion characteristics of a spark-ignition engine fueled with natural gas–hydrogen blends combining with EGR, International journal of hydrogen energy, **vol. 34**, p. 1035–1044 (2009)
5. T. Thurnheer, P. Soltic, S.I. engine fuelled with gasoline, methane and methane/ hydrogen blends: Heat release and loss analysis, International journal of hydrogen energy, **vol. 34**, p. 2494–2503, (2009)
6. C. Scholz, Koncepce jednoválcových motorů pro výzkum, XXXVIII. medzinárodná vedecká konferencia pracovníkov katedier a pracovísk spaľovacích motorov vysokých škôl na Slovensku a v Čechách. Bratislava, (2007)
7. C. Scholz, M. Fenkl, Pístový spalovací motor na vodík, Mezinárodní konference RT&E (2006)
8. M. Svoboda, M. Fenkl, Moving electrode spark plug experiments, XLI. International scientific conference of czech and slovak university departments and institutions dealing with research of combustion engines KOKA 2010, (2010)
9. M. Vojtíšek, M. Fenkl, Analysis of gaseous fuel blends and mixtures using inexpesive "garage-grade" gas analyzers., Journal of KONES Powertran and Transport 15. (2008)
10. V. Dvořák, Směšovací ventil, Patentový spis 298642, (2007)
11. C. Bleechmore, S. Brewster, Hydrogen Addition Strategy for Lean Limit Extension of a Natural Gas Engine, ANGVA 2007, (2007)

DOI: 10.1051/matecconf/20178901007

Laboratory detection of flat glass shapes using its reflection

Vlastimil Hotar[1,a], Ondrej Matusek[1] and Jan Svoboda[1]

[1]Technical University of Liberec, Department of Glass Producing Machines and Robotics, Studentská 2, 461 17 Liberec 1, Czech Rep.

Abstract. Glass is reflective in large angles of incidence. The using this property for detections of shapes is the basic goal of the research. The detection of 2D shapes from flat glass is a relatively simply example that has been used at the beginning of the research. The detection is based on three steps: capturing of an object in a large angles of incidence, inclusion distortion and other optical defects of the scan and a reconstruction of the shape.

1 Introduction

Vision systems for monitoring and control in the glass industry are applied especially for quality monitoring of container glass wares, window glass and automotive glass. The possibilities of the vision system applications in the glass industry are mostly: counting products, measurement and quality production monitoring, shape recognition, positioning and production monitoring and control using feedback. Despite an intensive investigation of machine vision for several decades, this is still a large field of research for solutions of some specific problems because glass has special properties [1]. The major problem is the transparency of colourless glass.

The systems have different requirement for monitoring glass melt and cold glass semi-finished or final products. The typical and general scheme of a monitoring and/or a control system is in Figure 1. Not all equipment mentioned in the figure must be used for an analysis of data from production.

The system generally consists of specific parts and for applications in glass production has some specific requirements:

- suitable tape of *illumination* must be chosen for scanning transparent materials (area array, backlight, ring illuminator, dome illuminator, diffuse light illuminator, multi-axis illuminator, polarized illuminator);
- *position of an object* must correspond with illumination;
- *adapters for lenses* can be used instead of (or with) illumination for better visibility of the object (polarisation illumination adapter, coaxial vertical illumination adapter, diffuse illuminator adapter, ...);
- *filters* can be used for the separation of some wave lengths from acquisition images;
- *lenses* (optic);

Figure 1. Diagram of monitoring and control systems

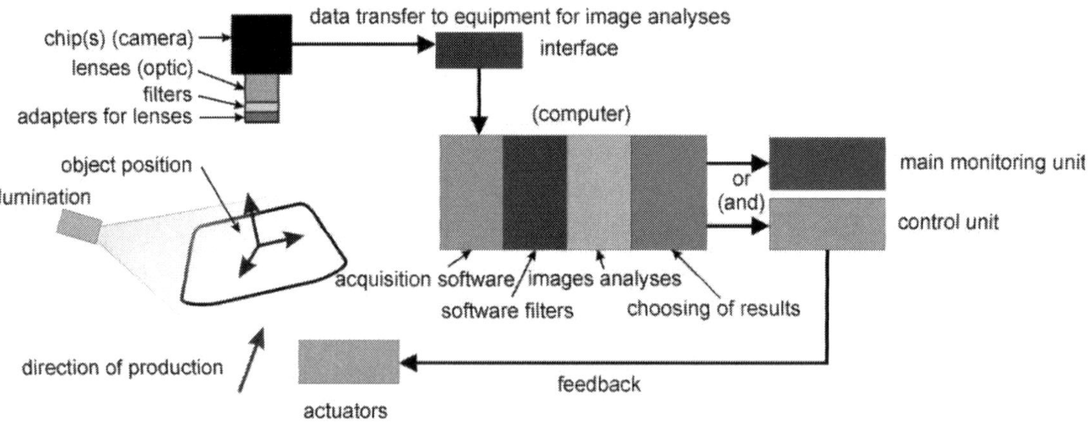

[a] Corresponding author: vlastimil.hotar@tul.cz

- *chip(s)* (camera), on the basis of the production process (line or area scan camera) and a required resolution (frame rate, resolution in pixels, colour, spectrum, ...);
- *data transfer to equipment for image analyses* (cables, A/D convector);
- *interface* (e.g. IEEE 1394, camera link, channel link, GigE vision, ...);
- *acquisition software*;
- *software filters*, important for the next process;
- *image analyses* [2], choosing fast and optimal analyses is important;
- *choosing results*;
- *data transfer to a main monitoring unit or a control unit* (connected by with feedback to actuators);
- *thermal insulation and cooling*, important in the case of glass melt monitoring.

The research at the Department of Glass Producing Machines and Robotics is focused to analyses of structured images using fractal dimension [3]. The data have character of digital images and obtained dividing lines (e.g. profiles [4], roughness and the dividing line of light and shadow). The analyses are applied for example in corrugation that uses a reflection of zebra plate [5]. The idea to use the reflection for a detection of glass objects comes from this application.

2 Glass reflection

When light moves from a medium of a given refractive index n_i (air) into a second medium with refractive index n_t (glass), both reflection and refraction of the light may occur [6]. In the Figure 2, an incident light ray PO strikes at point O the interface between two media of refractive indices n_i and n_t. Part of the ray is reflected as ray OQ and part refracted as ray OS. The angles that the incident reflected and refracted rays to the normal of the interface are given as θ_i, θ_r and θ_t. The relation between these angles is given by the law of reflection:

$$\theta_i = \theta_r; \tag{1}$$

and Snell's law:

$$\sin(\theta_i)/\sin(\theta_t) = n_t/n_i. \tag{2}$$

The fraction of the incident power that is reflected from the interface is given by the **reflectance** R and the fraction that is refracted is given by the **transmittance** T.

The amount of light reflected from the material under normal incidence (angle of incidence $\theta_i \approx \theta_t \approx 0$) is proportional to the square of the index change at the face:

$$R = [(n_i - n_t)/(n_i + n_t)]^2 \tag{3}$$

For common glass in air, $n_i = 1$ and $n_t = 1.5$; thus about = 4% of light is reflected. Note that reflection by a flat glass is from the front side as well as the back side, and that some of the light bounces back and forth a number of times between the two sides. The combined reflection coefficient R_g for this case is

$$R_g = 2R/(1 + R), \tag{4}$$

when interference can be neglected, R_g = 7.7 %. However, the reflection is for the intended detection insufficient, and illumination at a high angle must be used.

Figure 2. Scheme of reflection and refraction laws

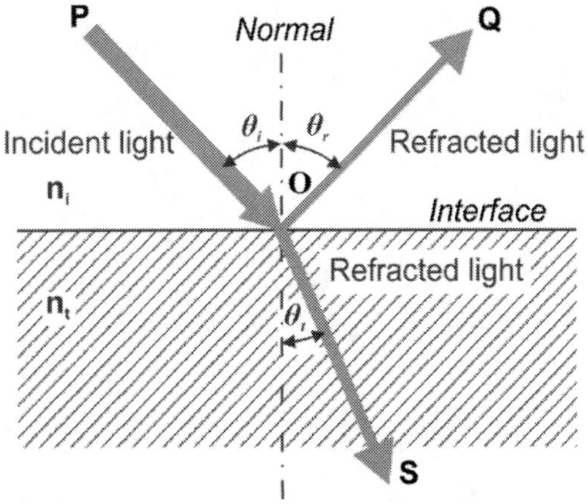

The calculations of R and T depend on polarisation of the incident ray. Using the Fresnel Equations (after simplifying) the equations for light polarized with the electric field of the light perpendicular to the plane of the diagram in Figure 2, the reflectance $R_\perp$ is given by:

$$R_\perp = \left(\frac{n_i \cos\theta_i - n_t \cos\theta_t}{n_i \cos\theta_i + n_t \cos\theta_t}\right)^2. \tag{5}$$

If the incident light is polarized in the plane of the diagram, the $R_\parallel$ is given by:

$$R_\parallel = \left(\frac{n_t \cos\theta_i - n_i \cos\theta_t}{n_i \cos\theta_t + n_t \cos\theta_i}\right)^2. \tag{6}$$

Using the Snell's law (2) the equations:

$$\theta_t = \sin^{-1}[(n_i/n_t)\sin\theta_i] \tag{7}$$

the $R_\perp$ and $R_\parallel$ are completely derived by θ_i. If the incident light is unpolarized (containing an equal mix of perpendicular and parallel polarizations), the reflection coefficient is

$$R = (R_\perp + R_\parallel)/2. \tag{8}$$

The dependence of reflectance on the angle of incidence is in the Figure 3.

As a consequence of the conservation of energy, the transmittance in each case is given by

$$T = 1 - R, \tag{9}$$

$$T_\perp = 1 - R_\perp \tag{10}$$

and

$$T_\parallel = 1 - R_\parallel. \tag{11}$$

 DOI: 10.1051/matecconf/20178901007

Figure 3. Dependence of reflectance on incidence angle and polarization of light

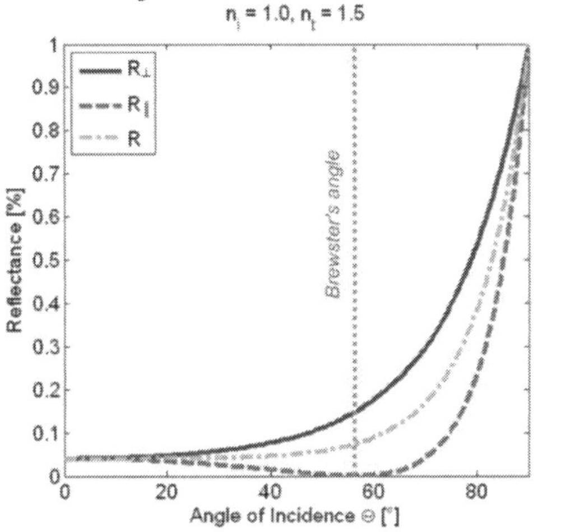

Figure 4. Dependence of reflectance on incidence angle for combined reflection coefficient R_g, unpolarized light

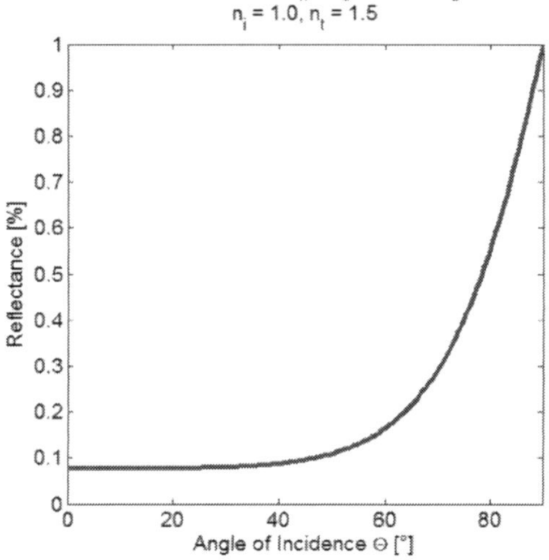

The Figure 3 shows that glass is reflective in bigger angles of incidence ($\theta_i \geq 70°$ for R and R⊥). Furthermore the reflectance depends on polarisation of incidence light; it is higher with the electric field of the light perpendicular. In the case of incident light is polarized in the plane (parallel), the reflectance $R_{\|}$ degreasing to the Brewster's angle, when the light is perfectly transmitted through the surface. Then the reflectance rises. The Brewster's angle is defined:

$$\theta_B = \tan^{-1}(n_t/n_i) \tag{12}$$

and for the plane glass surface $\theta_{Bg} = 56.31°$.

For the real measurement is the combined reflection coefficient R_g important, because of flat glass and reflection is from the front side and the back side. Using equation (4) the graph in Figure 4 is obtained for unpolarized light.

3 Detection of Flat Glass Shapes

The reflection under high angles of incidence can be used for detection of an object on a black, non-reflective and matt background. The angle of incidence should be selected on the basis of the function in the Figure 4. Obviously, the reflection does not dramatically change till angles of incidence 40° (8,6 % of reflectance, 7.7 % for the normal incidence). The reflectance is double for the angle 58,7° and triple for the angle 66,2°. If the higher reflectance under the high angle of incidence is used for detection of glass, the angle must be minimally 60°.

The reflectance can be increased using the perpendicular polarised lighting (Figure 3).

The detection is based on three steps, Figure 5:

1 capturing of an object in a large angles of incidence using a standard CCD or CMOS camera and a object-lens and obtaining of object contours,

2 recomputing of the captured object including distortion and other optical defects of the scan and

3 reconstruction of the captured shape including perspective.

Figure 5. Detection of flat glass shapes using reflection

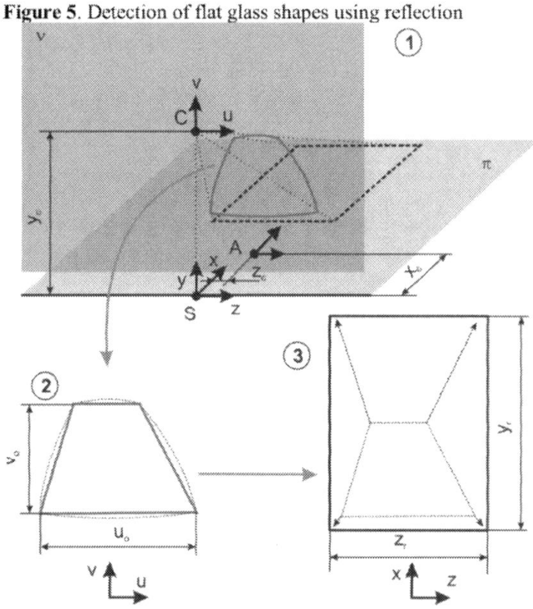

First two steps are commonly used in machine vision and image analysis. The third step has to be solved for the observed object. In the case of flat glass the one point perspective is used. Before starting the detection a pre-set

 DOI: 10.1051/matecconf/20178901007

must be done for the final detection. The per-set is a measurement with a ruler (an etalon), which defines relations (scale) between a captured space (in millimetres, axes x, z) and an image (in pixels, axes u, v). The position of the camera (the focal point of object-lends, x_c, y_c) and the axes start (A: x_c, z_c) must be also defined.

3.1. Experiment

Simple shapes were used in the experiment: circle (diameter 61 mm) and rectangle (length 50, width 40 mm). A chessboard with an edge of squares 5 mm was used for pre-set of the scale, Figure 6 and the basic relations for the conversion were defined. The position of the camera and the axes start were measured, Figure 7.

Figure 6 Chessboard image as etalon for determining of scale

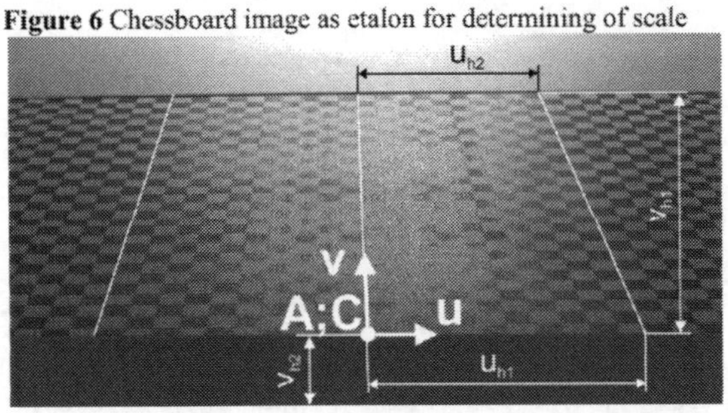

Figure 7 Position of camera and axes start

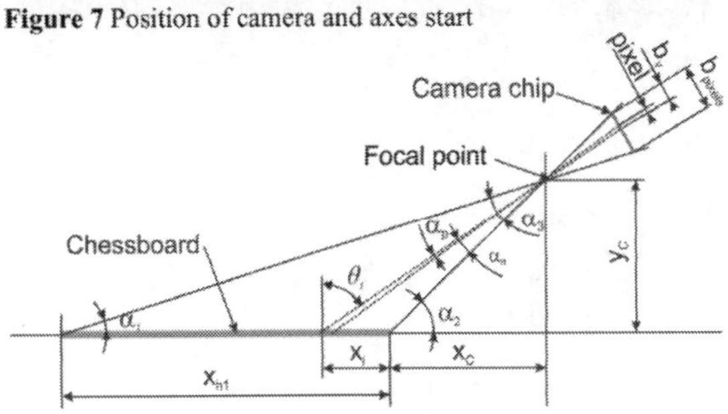

The functions for the reconstruction were derived, from the known measurement in pixels (axis u, v) to the real measurement in millimetres (axis x, z):

$$x_i = \frac{y_c}{\tan(\alpha_2 - (v_i \alpha_P))} - x_c \qquad (13)$$

where x_i is real position measured point in x axis in millimetres (axes start at the point A), v_i is position of the point in axis v at image in pixel (axes start at the point C, where A≡C), y_c and x_c are known, α_2 can be easily determined,

$$\alpha_p = \frac{\alpha_3}{v_{h1}} \qquad (14)$$

and

$$\alpha_3 = \alpha_2 - \alpha_1. \qquad (15)$$

For axis z can be determined the equation:

$$z_i = \left(\frac{u_i v_i}{h - v_i} + u_i\right)\frac{z_{h1}}{u_{h1}} \qquad (16)$$

where z_i is real position measured point in z axis in millimetres, u_i is position of the point in axis u at image in pixel, z_{h1} is length of the chessboard in millimetres corresponding to length u_{h1} measured in pixels, both are known, horizon is defined

$$h = \frac{u_{h1} v_{h1}}{u_{h1} - u_{h2}} \qquad (17)$$

where u_{h1}, u_{h2} and v_{h1} are obtained from the measurement from the chessboard image, Figure 5.

3.2. Results

Figure 8 and 9 show steps and results of the reconstruction.

Figure 8 Reconstruction of captured rectangle

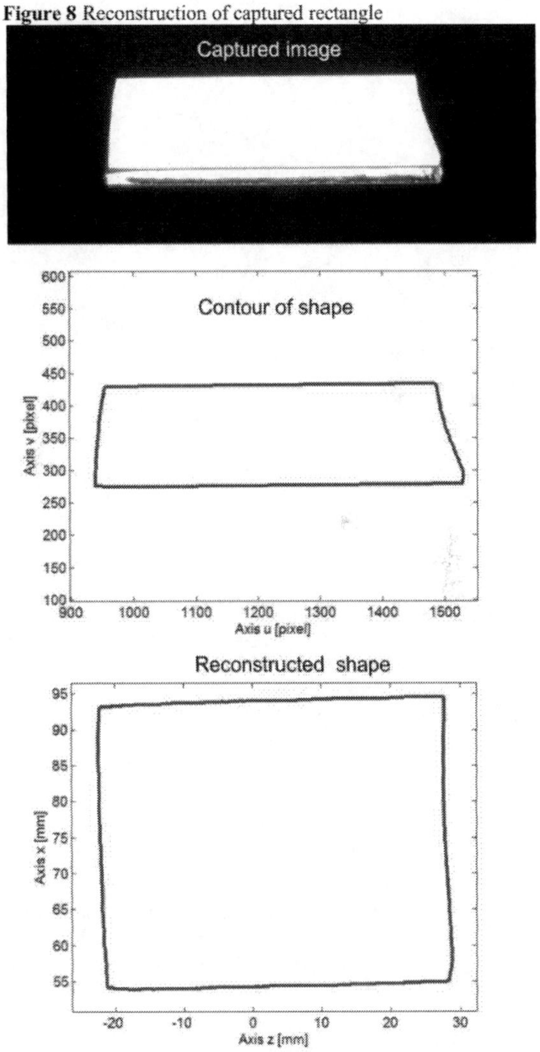

 DOI: 10.1051/matecconf/20178901007

Figure 9 Reconstruction of captured circle

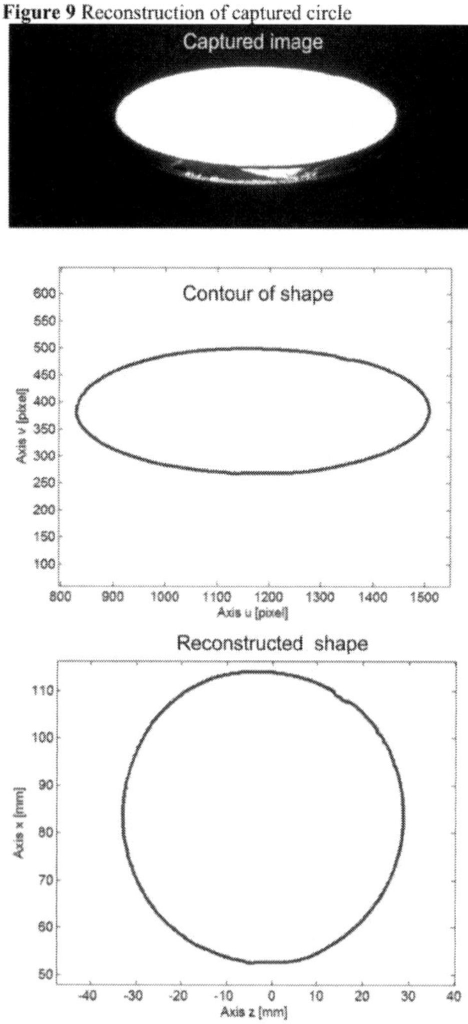

3.2 Discussion of results

The first analyses were focused to verify possibilities of the methodology and also to develop a software tool (in Matlab), hence the analyses were applied without the distortion of lens-object. The results are not perfect in accuracy and during the experiment problems were found that must be solved in next research, such as:

- accuracy in definition of the focal point,
- quality of the diffuse light,
- definition of the scale,
- inclusion of distortion and other optical defects.

In the next research effect of changed parameters has also to be defined, such as:

- polarization,
- intensity of a lighting,
- surrounding and parasite illumination,
- setting of a camera (with respect necessary depth of field),
- properties of a black, non-reflective and matt background,

dependence of the reconstruction accuracy on an angle of incidence.

4 Conclusion

The goal of this part of the research was to specify theoretical basic properties for the detection of glass using reflection under the high angles of incidence, to develop the software tool, verify possibilities of the methodology, and to specify problems and steps for next research.

The angle of incidence must be higher than 60°, optimal should be using the perpendicular polarised lighting, the equations for reconstruction were derived, the software tool was developed and was verified with experiment. The problems of the present approach and next the steps were specified.

The final target of the research is to specify conditions and a potential for the 3D analysis of formed flat glass like automotive one.

Acknowledgements

This work was supported by the grant of Students grant contest of the Technical University of Liberec, number SGS 21006/115, which use special-purpose support for the university research and is financed by the Ministry of Education, Czech Republic.

References

1. V. Hotař, *Monitoring of Glass Production Using Vision Systems*. In: 9th ESG Conference, Zurich: Trans Tech Publications Ltd. p. 511-516, (2008)
2. M. Sonka, V. Hlavac, R. Boyle, *Image Processing, Analysis, and Machine Vision*. Pacific Grove: Books/Cole Publishing Company, (1998)
3. B.B. Mandelbrot, *The fractal geometry of nature*, New York: W. H. Freeman and Co. (1982)
4. A. Hotař, P. Kratochvíl, V. Hotař, *The Corrosion Resistance of Fe3Al Based Iron Aluminides in Molten Glasses*. Kovové Materiály - Metallic Materials, p. 247-252, (2009)
5. V. Hotař, F. Novotný, H. Reinischová, *Objective evaluation of corrugation test*. Glass Technology: Eur J Glass Sci Technol Part A., Vol **52**, No 6, 197-202 (2011)
6. E. Hecht, *Optics*. 4th Edition. San Francisco: Pearson Education, Inc., (2002)

Laboratory analysis of the main properties and a color stability of a coating layers under the UV loading

Petr Kulhavý[1,a], Eva Kaniová[2], Vítězslav Fliegel[1] and Michal Vik [2]

[1]Technical University of Liberec, Dept. of Design of Machine Elements and Mechanisms, Studentská 2, 461 17, Liberec 1, Czech Rep.

[2]Technical University of Liberec, Dept. of Material Engineering, Studentská 2, 461 17, Liberec 1, Czech Rep.

Abstract. For individual components as well as for the final products is need to determine their resistance and other mechanical properties, which can be changing due to the various conditions and simultaneously we try to get still closer to the real operating conditions.Currently, given the need to guarantee a perfect quality of the product throughout its entire life cycle, it is necessary to assess the protective layers that are in direct contact with the material itself or with the future products.This article presents the results of testing individual components of the selected materials and their resistance to the UV rays. These were some samples of a prelacquered sheet and a powder coated plate after their forming, next by pigment colored concrete cubes, artificial wood and a prototype sample of the translucent material Licrete. Compared to a conventional climate tests such as heat, cold, humidity or the salt chambers, the colored protection layers were subjected to a strong UV rays. Subsequently was evaluated the change of the colors, brightness and where it was possible also the adhesion parameters and thickness of the coated layer. The measurement wasn't focused only on the adhesion and color stability, but also to a violation of microstructure due to high temperatures caused by the UV ray. The selected specimens were tested in a device which simulates solar radiation and then were them evaluated the changes of a colors and on the surface structure.

1 Introduction

Nowadays it is a considerable emphasis on to the appearance and reliable protective function of the surface layers placed on various ground materials. When choosing surface treatment and its subsequent application, we give into account the technological characteristics of the materials and also the characteristics of the used paints. Almost all of the products are always enclosed by the surface that is in contact with an ambient environment. Therefore is still there a continual mutual contact of metallic, non-metallic, liquid or the gaseous phases. Quite commonly for each of constructional materials we are looking for to its physical and mechanical properties with respect to the real mechanical and climatic loads. In the case of colored surface layers there are generally two aspects. Primarily we evaluate their properties such as the adhesion, abrasion resistance, heat resistance, chemical resistance etc. The second aspect to that we need to focus is an assessment of the surfaces exposed to the UV radiation and significance of its effect to the other parameters. The tested materials were a standard steel sheet and pre-lacquered steel plate Myriatouch. The used colored powders were from the manufacturer Sava trade and the manufacturer Beton Brož provided the prototype of concrete parts.

2 The surface layers

Surface treatments are primarily used for two reasons. It is an economic aspect, when we try to avoid or a significantly slow down the process of depreciation of the parts or components and also prolongs their life cycle.
The second aspect could be aesthetic, when the surface finish itself, used color and their properties or processing may become the one of the pillars on which is built the sale of the final product. The surface of material generally represents the interface on which is the gradient of the electrostatic potential associated with the charge a layer and there are some changing of a chemicals concentration. Due to an unbalance of the energy field on the surface of materials arises mainly for a metallic materials top surface tension and surface energy (the most densely occupied crystalline planes with the smallest surface energy). These physicochemical events have results in a gradual or complete disturbance of the surface layers. This process gradually affects almost all construction materials.

2.1 The tested materials

For a testing the surface structure and a color fastness of the materials were chosen samples made of steel,

[a]Corresponding author: petr.kulhavy@tul.cz

aluminum and concrete. Two types of sheet metal parts (thickness 1 mm) that have been created by a two different coating procedures.

The first sample was already pre-lacquered by supplier and under the protection film it has gone through the whole production cycle, while the second sample was powdered by epoxy just after the complete machining. The other tested samples were a concrete cubes with a by pigments tinted surface layers. Made from concrete is also an imitation of wood and the prototype of translucent material Licrete. The last samples are with various methods anodized aluminum profiles.

2.1.1 Steel

Two used types of steel samples were in colors white RAL 9010 and black RAL 9005. The used powder paint CPC-41-1 was a smooth semi-gloss, recommended for interior, from a series of hybrid powder coatings based on polyester-modified with epoxy resins. Polyester powder coatings are generally applied for high-gloss applications, or also where the UV stability (weather resistance) is needed. In Fig. 1 there is the distribution function of the used powder coat particles size. The value of the distribution function indicates show probability that the particles are smaller than the given size.

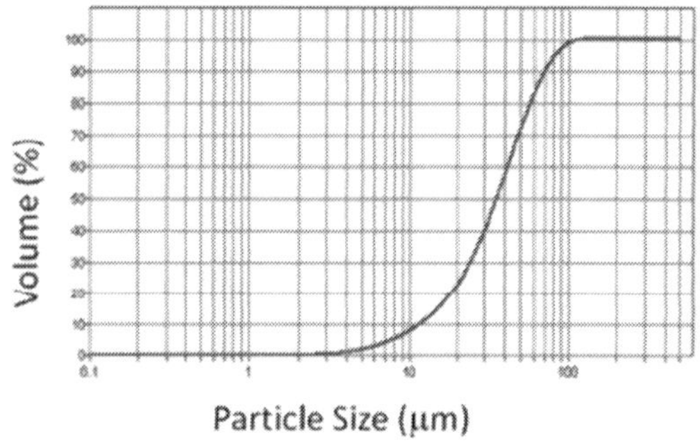

Figure 1. Particle sie distribution [1]

2.1.2 Concrete parts

For concrete components were used two samples. On the pigmented surface layers of the concrete cubes is controlled the surface color stability and for material Licrete (Fig.2) the rated luminous properties of the translucent parts. LiCrete® consisting of concrete and transparent elements that serve as a internal light guide. These were simultaneously occasion to check whether the action of UV worsens the translucent parameters and also eventually disrupt connections of the interfaces the two materials due to thermal expansion.

2.1.3 Anodized aluminum

These layers has been typically used for two reasons, the first is the possibility of industrial shading in an almost unlimited range of colors.

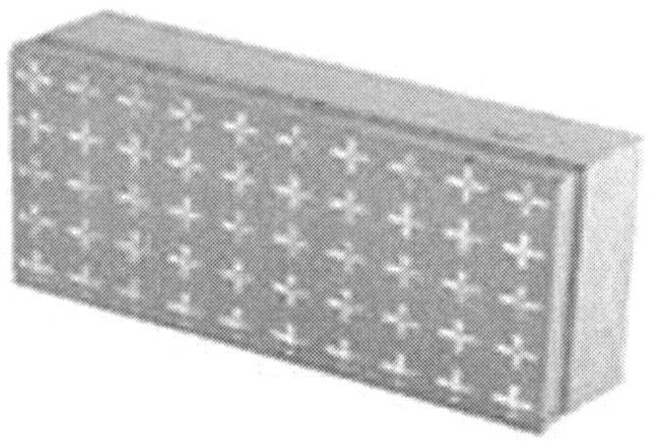

Figure 2. The translucent concrete "Licrete"

The second advantage is the resulting hard oxide layer of material surfaces, which provides better mechanical properties and chemical resistance. The thickness of the layers is usually in the range 5 - 20 μm. The tested color tones were Champagne (metal part in down section of Fig.3) and Silver (Fig.3 no.I standard, no. II sandblasted).

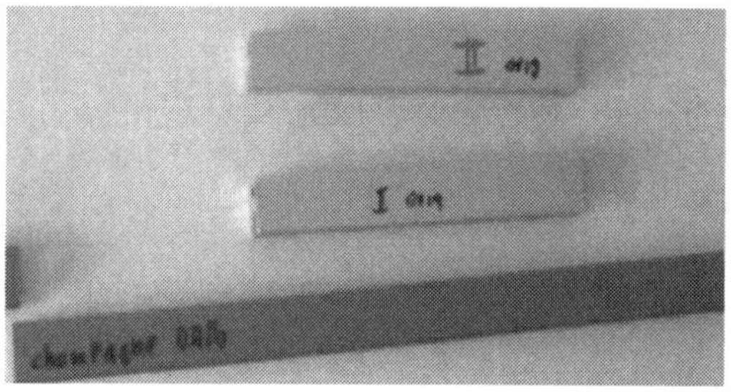

Figure 3. Aluminum profiles

2.2 The possibility of testingthe surface layers

For concrete materials are used primarily some typical tests according to internal standards of their manufacturers with regard to their specific application. For the anodized material is the test focused on the optical surface inspection, or it is also possible to check the closing of the pores on the layers with a hydroxide or acid.In contrast, for a metallic materials coated with powder paint we have by the standard [2] established tests:

- The electrical inductive measuring of the layer film thickness according to CSN EN ISO 2808, where some chosen individual points on the surface of the product defining the thickness of the surface layer [3].

- Macro images, visual tests used to detection defects under the surface, impurities, bubbles or cracks typically generated in response to the thermal expansion usually in a wrought locations.

- Adhesion tests of the surface layer, where is currently the one of the most popular called: *Grid test of the paint adhesion*. The principle of the test is to carve into the coating grid, which is, consists from orthogonal slices.

Each of the individual slices must pass through the paint to the surface of the main material.Then is the cutting line cleaned by brush and attached with the special adhesive tape, which is subsequently torn off. The quality of the coating is evaluated according to the degree of damage of the grid during the secession of the adhesive tape. The adhesionis according to [4] rated on ascale from 0 to 5.

3 About the UV test

The aim of this research was to compare the color change, resistance and other mechanical properties of coated layers exposed to operational use such as artificial light ageing in SUNTEST CPS+ device (Fig. 4).

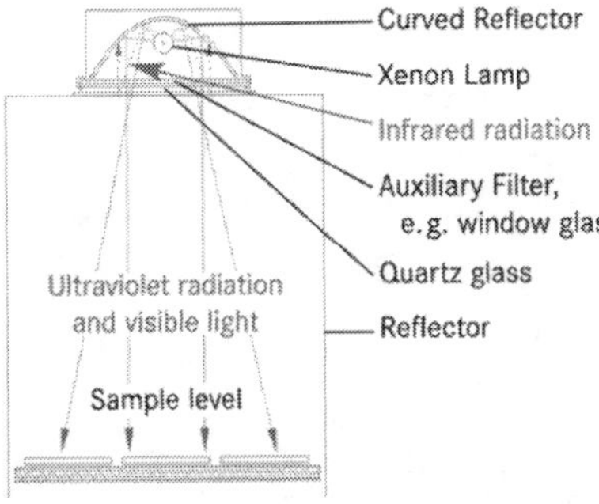

Figure 4. Test Chamber and Lighting System of SUNTEST CPS+ device

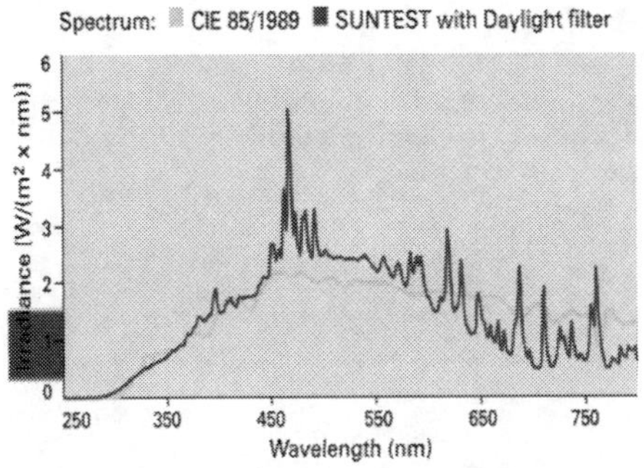

Figure 5. Composition of the used light

Artificial light ageing tests has been carried out according to ISO standard [5] in a Suntest CPS+ device. The following measuring conditions were applied:

- irradiance: 662 W/m^2,
- black standard temperature:100°C,
- daylight filter (lower limit — 290 nanometer).

The filter used allowed to simulate the solar radiation present outwards. Ten different samples were tested to light exposure for 72 hours, which is almost equal to 34 days of ageing under natural outdoor conditions and about 42 days of ageing under natural conditions prevailing indoors. The dimensions of the examined samples were different based on preparation method. After every 24-hour exposition to light in Suntest CPS+ device the papers were evaluated visually, then the optical properties were measured.

3.1 Measurement of colorimetric properties

The properties such as color coordinates CIELAB as well as color attributes CIE LCh$_{ab}$ were measured in first step under the following conditions:

- illuminant D65,
- 10° CIE standard observer,
- viewing condition di:8°,

by Datacolor SF600+ spectrophotometer, using following tristimulus values *XYZ* (*T*) calculation:

$$X = k\int_{\lambda} E_{\lambda} R_{\lambda} \bar{x}_{\lambda}\, d\lambda \tag{1}$$

$$Y = k\int_{\lambda} E_{\lambda} R_{\lambda} \bar{y}_{\lambda}\, d\lambda \tag{2}$$

$$Z = k\int_{\lambda} E_{\lambda} R_{\lambda} \bar{z}_{\lambda}\, d\lambda \tag{3}$$

$$k = 100 \Big/ \int_{\lambda} E_{\lambda} \bar{y}_{\lambda}\, d\lambda \tag{4}$$

from which were computed L*a*b* values in case of ratios of tristimulus values (T/T_0) > 0.008856:

$$L^* = 116\sqrt[3]{\left(\frac{Y}{Y_0}\right)} - 16 \tag{5}$$

$$a^* = 500\left[\sqrt[3]{\left(\frac{X}{X_0}\right)} - \sqrt[3]{\left(\frac{Y}{Y_0}\right)}\right] \tag{6}$$

$$b^* = 200\left[\sqrt[3]{\left(\frac{Y}{Y_0}\right)} - \sqrt[3]{\left(\frac{Z}{Z_0}\right)}\right] \tag{7}$$

Where: T_0 are tristimulus values of perfect reflecting diffuser at used illuminant (D65: X= 94.811; Y = 100 and Z= 107.305). In other cases, when ratio of tristimulus values is less or equal then 0.008856, are used linear transformation functions [6].

The L*a*b* system is also referred to as the CIELAB system. It can be visualized as a cylindrical coordinate system in which the axis of the cylinder is the lightness variable L^*, ranging from 0 to 100, the radius is the chromaticity variable C^*, and hue angle $h°$:

$$C^* = \sqrt{(a^*)^2 + (b^*)^2} \tag{8}$$

$$h^{\circ} = \arctan\left(\frac{b^*}{a^*}\right) \text{in degrees (0-360°)} \tag{9}$$

As third step was calculated color difference ΔE* between specimen data after and before exposure:

$$\Delta E^* = \sqrt{(\Delta L^*)^2 + (\Delta a^*)^2 + (\Delta b^*)^2} \quad (10)$$

Due to high roughness of some samples was used substantial method based on non-contact measurement. Each sample was firstly captured by digital camera CANON EOS300D in viewing box SPL III (X-Rite) under daylight simulator D65 and viewing conditions 45°x:0° together with 24 color standards of X-Rite ColorChecker (CC24) as visible on Fig. 6:

Figure 6. Image of the concrete sample with CC24

As last step of this procedure were measured RGB values of pixels relating to sample and color standard CC24, which were transformed into XYZ values via the following transformation matrix **B** computed from known tristimulus values of CC24 color standard by optimization procedure:

$$\begin{pmatrix} X \\ Y \\ Z \end{pmatrix} = B^{-1} \begin{pmatrix} R \\ G \\ B \end{pmatrix} \quad (11)$$

4 Test results

A visual evaluation showed that after third cycle of exposure are visible color differences between samples on the start and the end of experiment and for people with superior color discrimination for all of the tested samples. All observers agreed that most visible difference was assessed for sample imitating wood.

Due to the mentioned problems with contact measurement of color coordinates by spectrophotometer are there presented only color coordinates based on non-contact measurement. Tab. 1 and Tab. 2 shows L^*a^* b^* of the samples before exposure to UV light and after 72-hour of exposure.

Table 1. Color coordinates before exposure (on the start)

Sample	unexposed					
	L*	a*	b*	C*ab	hrad	hab
Steel RAL 9005 černý	24.68	0.13	-0.43	0.45	5.00	286.73
Steel Ral 9010 bílý	93.57	-0.72	4.54	4.60	1.73	98.99
Imitation of the wood	67.45	1.05	12.77	12.81	1.49	85.32
Licrete (concrete)	82.62	-1.37	7.32	7.45	1.76	100.57
Al Champagne	67.46	1.08	12.89	12.94	1.49	85.22
Al Silver	85.66	-0.86	1.52	1.75	2.08	119.43
Al Silver Corado	86.36	-1.03	1.06	1.48	2.34	134.36
Concrete cube red	62.73	10.41	14.49	17.84	0.95	54.30
Concrete cube yellow	67.55	6.77	26.87	27.71	1.32	75.86

Table 2. Color coordinates after 72-hour exposure

exposed					
L*	a*	b*	C*ab	hrad	hab
25.23	0.10	-0.63	0.64	4.87	279.21
93.05	-1.05	4.67	4.78	1.79	102.73
60.74	5.06	11.48	12.55	1.16	66.23
86.55	-1.44	5.55	5.73	1.82	104.51
64.90	1.26	14.41	14.47	1.48	85.00
85.81	-0.94	0.93	1.32	2.36	135.16
86.82	-1.06	0.43	1.15	2.76	158.21
59.29	12.30	15.37	19.68	0.90	51.33
72.69	6.92	28.49	29.32	1.33	76.35

Color differences between samples on the start and on the end after third cycle of exposure are represented in Tab. 3 together with graph on figure 7, where is visible effect of light exposure on differences in individual coordinates.

Table 3. Color differences after 72-hour exposure

difference					
dL*	da*	db*	dC*	dH76	dE*
0.55	-0.03	-0.21	0.20	-0.07	0.59
-0.53	-0.34	0.12	0.19	0.31	0.64
-6.71	4.01	-1.28	-0.26	-4.20	7.93
3.93	-0.07	-1.77	-1.71	0.45	4.31
-2.56	0.18	1.52	1.53	-0.05	2.98
0.15	-0.08	-0.59	-0.43	0.42	0.61
0.47	-0.03	-0.63	-0.33	0.54	0.79
-3.44	1.89	0.88	1.85	-0.97	4.02
5.13	0.15	1.62	1.60	0.24	5.38

Graph on the figure 7 show that samples with measured high color difference (above 4 units) are affected by change in lightness L^*, where only concrete samples show increasing of this value. In other cases was visible decreasing of lightness value. This effect is mainly caused by increasing of roughness of tested samples as visible on picture in figure 10.

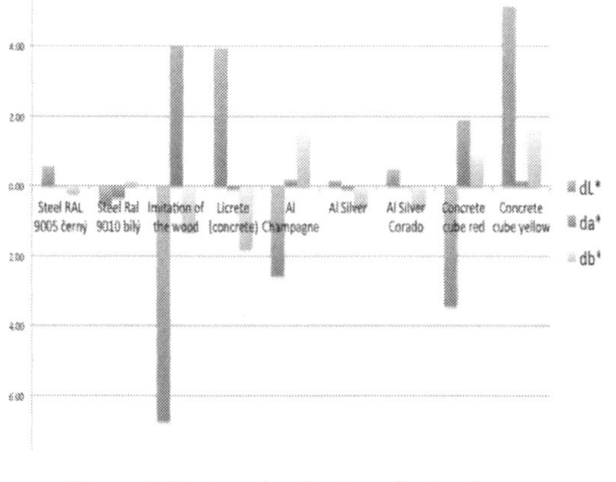

Figure 7. Vectors visualization of color changes

Beside effect of exposure on lightness values as second most visible effect is found for color coordinate b^*, which is the blue (negative) to yellow (positive) axis. There are only two exceptions for sample imitating wood and concrete cube red, where is visible color change measured in color coordinate a^*, which is the green (negative) to red (positive) axis. It is interesting; that in both situations measured effect was positive color change, because expecting effect is decreasing of chromacity, which means sample after exposure is obviously duller than on the start. Possible explanation of this result is artificial simulation of sunlight exposure, which is in natural conditions obviously followed by dust exposure. Such improvement of test procedure is now prepared as continuous research in weathering testing of materials used in civil engineering.

In figure 8 are visible color differences on the chromatic plane of CIELAB color space, where are showed samples before (circle) and after third cycle of exposure (triangle).

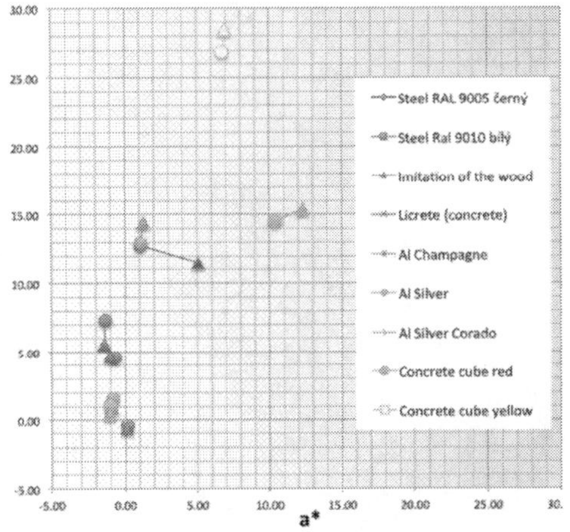

Figure 8. Vectors visualization of color changes

5 Measurement of physical properties

For the concrete materials the fundamental to us is primarily the color stability of the pigment layers. For painted and anodized parts components is need to consider some more properties. This is an assessment of the effect of UV radiation to the structure, thickness of lacquer layer, surface quality and possibly also changes in the adhesion properties.

5.1 Determination of the dry film thickness

Using the device Salutroncombi D3M has been carried out measuring of the thickness of the surface layer. This device works generally on the principle to receiving electromagnetic waves reflected from the metal surface. For materials based on Fe is used anelectromagnetic induction and for the non-ferrous and nonconductive materials are used whirlpool currents.

The device used for measuring the thickness of the layer is before measuring inspected and calibrated with using an etalon foil. The layer is always measured at several points of the product, ideally in the middle and on the edge regions if it is possible. Simultaneously are evaluated some frequently occurring defects under the surface. Typically impurities, air bubbles or cracks caused generally in response to the thermal dilatations

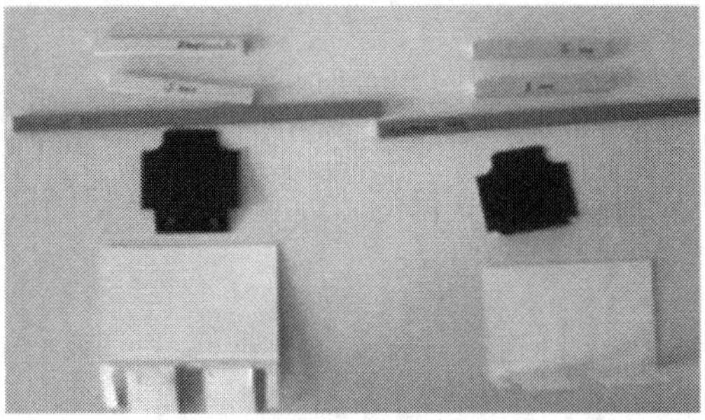

Figure 9. Measured Fe and NonFe parts

Table 4. Thickness of the coat layer (µm)

	Unexposed		Exposed	
	Mean value	St. deviation	Mean value	St. deviation
Pre-lacquered sheet	45,5	2,94	42,1	4,04
Powder coated sheet	50,1	2,9	52	5,04
AL Champaine	18,1	2,67	21,25	2,9
AL Silver	19,7	1,82	22,15	0,68
AL Silver Corado	18,5	4,34	16,2	1,7

Due to big differences of the thickness of each part, it's not possible to accurately assess the potential influencing of the UV ray. Therefore according to Tab. 4 there can't be observed a provable influence to the resultant thickness of the layer.

5.2 Grid test of the paint adhesion

On all powder coated parts, was also carried the most important test, indicating the value of adhesion. Into a series of samples were before and after illumination with using the standardized incisor cut grid (Fig.10) from which was subsequently pulled down the special adhesive ISO tape. After the tearing of the tape has been under zooming evaluated the resulting deformation and vacancies of the grid.

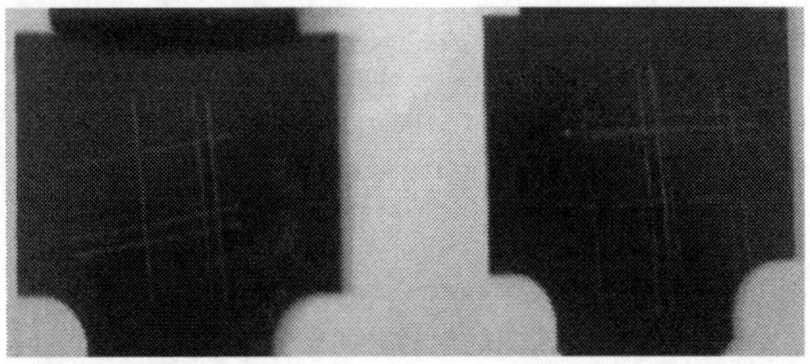

Figure 10. The grid after application of the adhesive tape

5.3 Macro-surface structure

Due to thermal dilatation, especially in the case of materials combined of several elements, it was necessary to assess also their macrostructure. The structure is usually evaluated by the microcamera under good lighting conditions. In case of finding defects there are typically the cracks, bubbles or separation between phases. On the measured samples any similar cases has not been detected. It can be stated that the UV ray and temperatures around 100 °C don't threaten the surfaces.

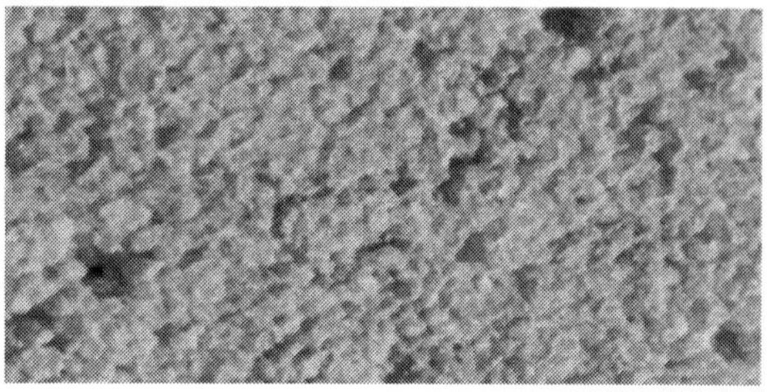

Figure 11. An example of the microstructure

6. Conclusion

We can say that there hasn't been found any demonstrable influence of the UV rays or the simultaneously generated heat to any of the observed mechanically measurable parameters. It has been found out that the only changes were in the tones and the brightness of the colors coating on the surfaces. Recognize the color change with the naked eye, it was possible only for one of the materials, namely the white powder coated steel. Find the differences in the other cases were possible only with using a specific computer optical viewing technology.

Acknowledgements

This work was supported by the project TKMOST.

References

1. Technical data sheets of the specialized powder coatings manufacturer Sava trade ltd.
2. ČSN EN ISO 4287. *Geometrické požadavky na výrobky (GPS) – Struktura povrchu: Profilová metoda. Termíny, definice a parametry struktury povrchu.*Praha: Český normalizační institut, 1999.
3. ČSN EN ISO 2808. *Nátěrové hmoty – Stanovení tloušťky nátěru.* Praha: Český normalizační institut, 2000.
4. ČSN ISO 2409. Nátěrové hmoty – Mřížková zkouška. Praha: Úřad pro normalizaci aměření, 1992.
5. ISO 11341:2004: *Paints and varnishes — artificial weathering and exposure to artificial radiation — exposure to filtered xenon-arc radiation.*
6. CIE 15:2004 *Colorimetry*
7. MAZUREK, Pavel. *Práškové barvy a jejich odlupování.* MM Průmyslovéspektrum. 2005.
8. MAHMOUD, Manal G. Influence of ultraviolet light irradiation on corrosion behavior of weathering steel. *Scripta Materialia.* 2005, vol. 53, DOI: 10.1016/j.scriptamat.2005.07.039.
9. http://www.casopis-koroze.cz/index.php

Measuring frequency and transient characteristics on Yaskawa Motor and controlling the motor by means of PC

Martin Lachman[1,a]

[1]Technical University of Liberec, Department of Manufacturing Systems and Automation, Studentská 2, 461 17 Liberec 1, Czech Rep.

Abstract. This article familiarise its readers with possibilities of motor control, measuring frequency and transient characteristics on Yaskawa Motor in laboratory conditions using Matlab – Simulink software and its Real Time Target toolbox. Using the Real Time Target toolbox along with MF624 I/O card, we are able to capture, store and submit information on state variables of the motor in real time. The manufacturer of Yaskawa motors does not provide with information about frequency and transient characteristics of current, speed or position loops. In order to verify the mathematical model, it is necessary to measure these characteristics on a specific motor type and use the obtained results to recalculate amplification of individual controllers by means of the mathematical model. Adjusting the current controller is not possible, since the values are fixed by the manufacturer. The manufacturer allows the user of the current controller to choose from 10 pre-set values (from the „softest" to the „most stiff"). By making the model more accurate, we may perform simulations of various load states that are approaching real ones.

1 Introduction

Parameters of the motor mathematical models were obtained from the motor manufacturer's documentation and from measuring reaction of required torque jump when connected to a torque (current) feedback circuit (Fig. 1) and from reaction of required speed jump when connected to a speed feedback circuit. Even though the measurement and simulation processes on Figs. 1 and 2 corresponds very well with each other (reaction of torque and speed jumps), it was decided to perform the measurement directly on the motor in order to specify the shape of the frequency characteristics.

The measurement of frequency characteristics in speed feedback are shown in Chapter 3. A difference between measurement and simulation is expected mainly during higher frequencies. The measurement will also verify setting values of the speed controller, since the manufacturer only provides with time constant of the speed controller in seconds along with information about „indeterminate" amplification, whose unit does not make the amount of amplification apparent (from the physical point of view).

Fig. 1 Reaction of the assembly to the torque jump

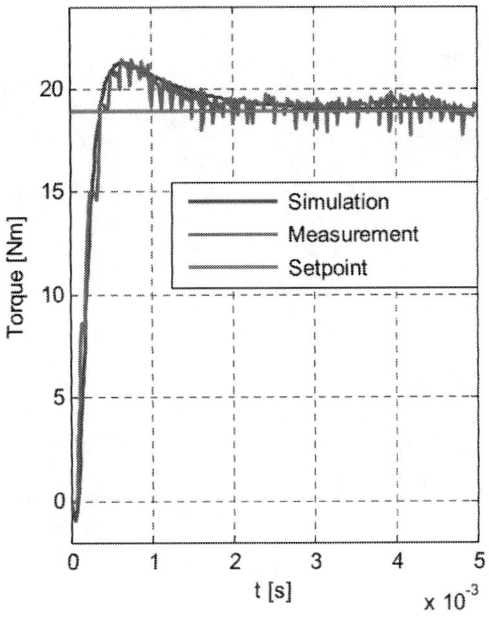

[a] Corresponding author: martin.lachman@tul.cz

Fig. 2 Reaction of the assembly to the speed jump

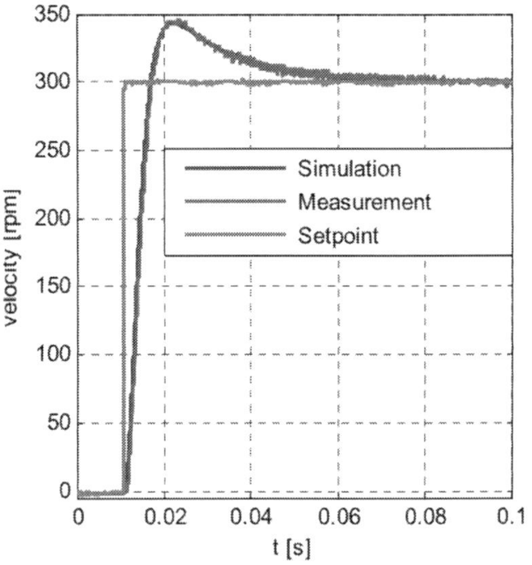

A PC with Matlab – Simulink software and Real – Time Windows Target toolbox installed was used for measurement along with Humusoft MF624 measurement card. The Real – Time Windows Target toolbox provides with communication between the MF624 measurement card and Simulink model.

2 Preparation of Frequency Characteristics Measurement

•Creating a Simulink model with marked inputs and outputs to MF624 card (Fig. 3) and set signal sampling period.
•In Real – Time Workshop (tab in Simulation – Configuration Parameters menu), we choose the target platform for Simulink model compilation. In our case, the platform is Real – Time Windows Target. Compilation of the model into C programming language – this part is performed by Simulink automatically.
•When the compilation is complete, we switch Simulink to External mode and press the Connect to Target button. The code compiled to C programming language is executed, establishing connection with the Real – Time kernel. That also results in measuring card initialisation and the parameters from the compiled model are set on the card as well.
The last step is to press the Start Real – Time Code button (same button we used for starting regular Simulink simulation) and we may start with the measurement.

Fig. 3 A simple Simulink model

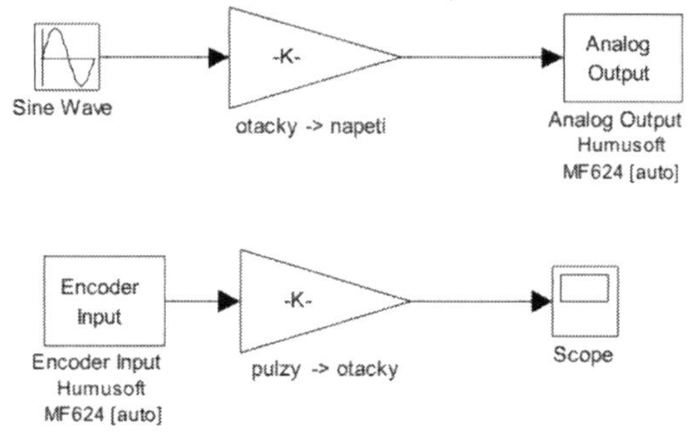

3 Possibilities of Performing Frequency Characteristics Measurement

1. Feed a pink noise to the speed loop input as the required speed value and measure the speed. A course of the amplitude and phase frequency characteristics may be obtained by means of mathematical apparatus. Unfortunately, this procedure did not turn out to be the right choice, since the output (analog) signal was severely deformed by the environmental influences, and it was not possible to use it as a credible result.
2. Feed a function of $A_1 \cdot \sin(\omega \cdot t)$ with A_1 defined amplitude and *f* frequency corresponding to $\omega = 2 \cdot \pi \cdot f$ to the speed loop input as the required speed value.

The actual speed of motor shaft rotation with A_2 amplitude and φ phase shift would be measured at the analog output. However, this method may be used only for frequencies of up to appx. 200 Hz; exceeding this value will lead to significant declination of A_2 amplitude and the noise modulated on the carrier signal will overcome the carries signal by its amplitude, resulting in erroneous evaluation of the measurement.
3. Feed a function of $A_1 \cdot \sin(\omega \cdot t)$ with A_1 defined amplitude and *f* frequency corresponding to $\omega = 2 \cdot \pi \cdot f$ to a speed loop input as the required speed value.

Unlike in the previous item, in this case, the analog speed will not be measured. What we will measure, however, will be the position pulses from the incremental sensor (turning of the shaft) and numerically derive them afterwards, thus providing us with information on actual motor shaft rotation speed.

This method did not lead to satisfying results as well, since the biggest problem was in the numerical derivation.
4. Feed a function of $A_1 \cdot \sin(\omega \cdot t)$ with A_1 defined amplitude and *f* frequency corresponding to $\omega = 2 \cdot \pi \cdot f$ to a speed loop input as the required speed value.

The difference is that in this case, we measured position (pulses), which we did not transform to speed by means of derivation.

By using correlation of input and output and respecting that the input is $A_1 \cdot \sin(\omega \cdot t)$ function and the output is $A_2 \cdot \sin(\omega \cdot t + \varphi)$ function, we are able to evaluate the frequency characteristics of the speed loop even on relatively high frequencies, where the output amplitude is becoming very low.

4 Comparison of Measurement and Simulation

Fig. 4 shows the comparison of measured and simulated amplitude and phase frequency characteristics of Yaskawa motor when in speed feedback circuit. The controllers were set to factory settings.

A very good conformity (mainly in case of amplitude characteristics) between measurement and simulation on mathematical model has been reached. Due to that, it may be concluded that constant values of speed controller in the mathematical models were appropriately chosen pursuant to speed jump reactions.

Fig. 4 Amplitude characteristics and phase frequency characteristics

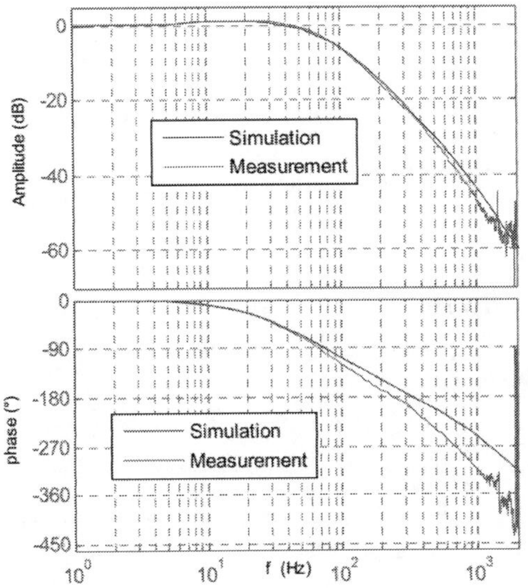

5 Basic Control Modes of the Yaskawa Inverter

5.1 Torque Control Mode

This is the most subordinate control mode. Its equivalent is the simulation scheme, where the input value is required current and the output value is the actual current. In this case, the required value is the required torque fed as analog voltage corresponding to the moment to CN1 connector, PIN 9. The output is actual torque of the motor shaft.

In case we want to control the turning of the motor shaft fitted with inverter in torque control mode, we shall use the higher-level control system with speed and position feedback. For this purpose, we would use a control computer (see Fig. 5) with Matlab software and Real Time Windows Target toolbox, as partially described in Chapters 2 and 3. The Pn400 constant contains conversion of voltage to motor torque.

Fig. 5 Connection of PC to the Yaskawa inverter in the torque control mode

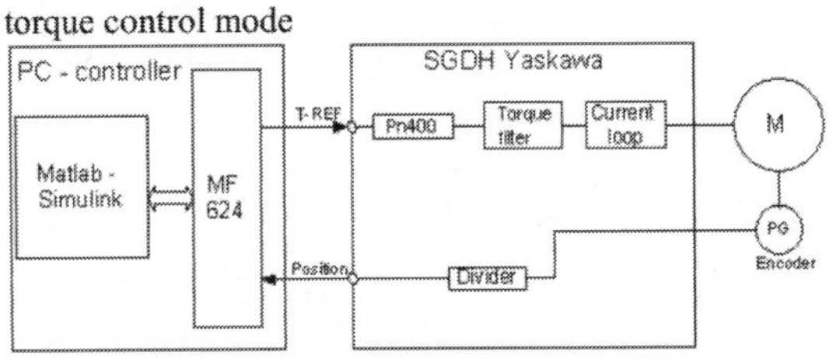

Fig. 6 PC - controller diagram in Real Time toolbox

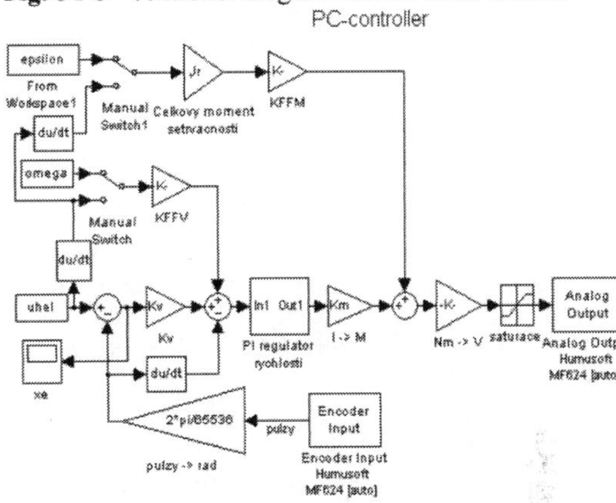

Fig. 6 shows two feedbacks (speed and position) and speed and torque feedforwards.

The encoder feeds information on current pulse position into the diagram. The information is converted to rad and is fed to the position loop as an actual position value. The difference between the required and actual position is multiplied by Kv constant (used to obtain required speed); the speed feedforward is consequently added to the value.

Feedforwards offer choice between position derivations, or we may calculate the course of the required speed and acceleration in advance and feed it separately to the schematic. „Manual Switch“ is used for that.

Advantages and Disadvantages of the Torque Control Mode

+ Possibility to create own control structures in PC, we are not limited by the fixed structure in the controller,
+ easy to add various filters,
+ possibility to easily choose data source for feedforward
- Calculation of speed derivations by means of decreased amount of pulses is not one of the most suitable,
- analog value of the required torque may be loaded with noise,
- high PC system requirements due to the necessity to use higher sample frequencies.

5.2 Speed Control Mode

In this control mode, the current (torque) feedback and speed feedback is closed in the Yaskawa inverter, while only the position feedback is closed in the higher-level control system. To provide a better understanding, the Fig. 7 contains constant values changed on the inverter.

The analog voltage value corresponding to the required speed (in rpm) is fed to the inverter via CN1 connector, PIN 5 (V-REF). We change the Pn002 constant in the inverter in order to be able to feed the analog torque feedforward value (T-REF) to CN1 connector, PIN 9.

Fig. 7 Connection of PC to the Yaskawa inverter in the speed control mode

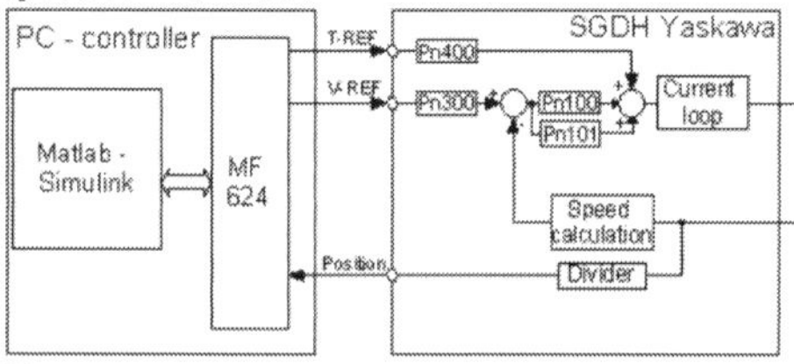

The Fig. 8 does not include the inverter filters. The inverter input is fed with both analog signals (T-REF, V-REF) through fixed filter with a time constant. In addition, there are 2 filter on the current loop input. The narrow band filter is set by the Pn409 constant, while the lowpass filter is set by the Pn401 constant. A filter set by the Pn307 constant is placed in front of the speed loop input.

The Pn100 and Pn101 constants are used to set values of speed PI controller. The Pn300 constant contains conversion of voltage to motor speed (rpm).

Fig. 8 Schematic of PC controller for speed control mode

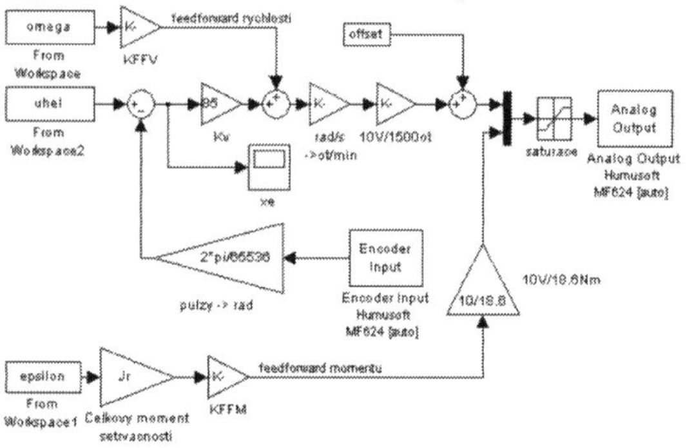

Advantages and Disadvantages of Speed Control Mode

+ Possibility to create own controller structures in PC
+ easy to add various filters,
+ possibility to easily choose data source for feedforward.
- analog value of the required speed may be loaded with noise, consequently affecting the xe deviation position. An offset compensation shall be performed by means of offset.

5.3 Position Control Mode

In this mode, all 3 controller loops (current, speed and position) are closed in the inverter and the higher-level control system feeds the inverter only with information on required position; the output of the inverter is the actual position of motor shaft (in pulses as well).

Advantages and Disadvantages of Position Control Mode

+ Noise no longer affects the signal between PC controller and inverter (unless analog feedforward is used).
- The controller fixed structure cannot be modified,
- it is crucial to use accurate path-generating pulse generator.

5 Conclusion

It is worth noting that when using method No. 4 described in Chapter 3, we were able to perform the measurement above the frequency of 1,000 Hz. Around the frequency values of 1,500 Hz, there is a noticeable resonance peak on the amplitude characteristics.

This matter is of mechanical nature and is probably caused by vibrations of the shaft sensor (due to sensor mounting). The resonant frequency was measured several times in the speed feedback (to exclude measurement errors), even with various controller setting. However, the amplitude peak occurs repeatedly.

Research in this article was supported by targeted support for specific university research in terms of the TUL student grant competition (Project 21010 – Complex Optimisation of Manufacturing Systems and Processes 2).

References

1. J. Skalla, Dynamické chyby dráhy při kruhové interpolaci NC obráběcích strojů, Liberec (2003)
2. M. Lachman, Regulace přesných polohových servopohonů při vysokých rychlostech, disertační práce, Liberec, (2005)
3. P. Souček, Vysoce dynamické pohony posuvů obráběcích strojů, Praha (2002)
4. R. Mendřický, Pasivní odpory krytů vedení obráběcího stroje. In: Sborník Vědecká pojednání XIV / 2008 - ACC JOURNAL, Technická univerzita v Liberci, Liberec 2008, ISBN 978-80-7372-379-8, ISSN 1801-1128.
5. R. Mendřický, Improved Friction Models for Feed Drives. In Machine tools, Automation and robotics in mechanical engineering, MATAR PRAHA 2004, ISBN 80-903421-3-2.

 DOI: 10.1051/matecconf/20178901010

Hexapod control system and software

Aleš Lufinka[1,a]

[1]Technical University of Liberec, Depaartament of Design of Machine Elements & Mechanism, Studentská 2, 461 17 Liberec 1, Czech Rep.

Abstract. Hexapod is a testing device for generation of six degrees of freedom movement. Its control system is built from the National Instruments components and control application is created in the software LabVIEW. Using the real-time system to run this application requires special programming procedures because the precise timing of the control loop has to be observed. The hexapod motion control is based on its kinematic model. Equations for the six basic motions (3-axis movement and 3-axis rotation) were obtained from the kinematic model. These equations are implemented directly in the control program for basic movement and rotation control. The control system, program design and program optimization are described in this paper.

1 Introduction

The hexapod is a new testing device for generation of six degrees of freedom movement (see Fig.1). It is used for testing of mechanical parts or assemblies properties (mostly car seats).

Figure 1. The hexapod testing device.

This kinematical solutions increased device stiffness so higher frequencies of testing signals can be used. Of course the control signals calculation is more complicated than for the classical orthogonal system. Hexapod motion is induced by six linear hydraulic engines which are controlled by the position feedback. The control system is built from the National Instruments components and control application is created in the software LabVIEW. Because the precise timing of the control loop has to be observed the real-time system to run the control application has to be used. The hexapod motion control is based on its kinematic model, equations for the six basic motions (3-axis movement and 3-axis rotation) were obtained from this model. These equations are implemented directly in the control program for basic movement and rotation control.

2 Control system hardware

The control system block diagram is shown in the next figure.

Figure 2. The control system block diagram.

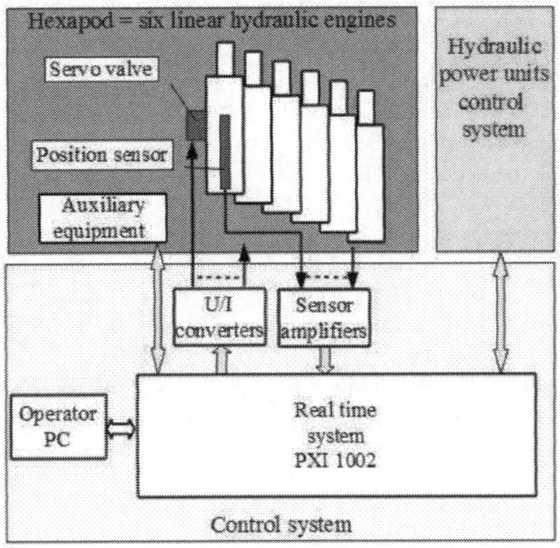

[a] ales.lufinka@tul.cz

National Instruments real-time system PXI1002 forms the control system core [1]. Piston position sensors are connected to the six channel amplifier and then to the PXI system by 16-bit A/D converter. Voltage signals from PXI output 16-bit D/A converters are transferred to the current signals for the servo valves control. The above elements constitute the six hydraulic engine basic position feedback.

Galvanically isolated digital inputs and outputs control all hexapod auxiliary equipment (input hydraulic valves, signal lights, etc.). Digital lines are used for communication with the laboratory hydraulic power units control system [2], too. Direct hydraulic power units control is thus possible by the hexapod software application.

Figure 3. The hexapod control system.

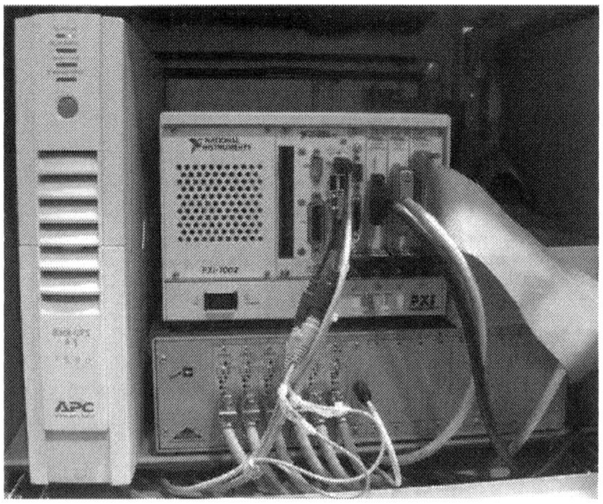

The standard PC provides user-friendly interface for the hexapod control. This PC and real-time system are connected by the classic internet cable.

3 Hexapod control software

The hexapod control software was created in LabVIEW 2009 RT. A LabVIEW real time project defines a set of applications and their distribution between the RT system and PC. It also defines the priority policy and the data exchange system. In this case, applications are mainly allocated to the real-time system. Only the user interface front panel and pre-processor application are placed to the user PC.

3.1. Real-time application structure

Three parallel running loops form the core of the most important real-time application. These loops are very accurately timed by the RT system 1Mhz hardware clock and their run priority is precisely defined. Six PID controllers are realized by the high priority hardware timed (2kHz frequency) loop. Exact running of this loop must not be breached in any case so all data are transmitted to this loop by a data queue. Desired position values are read from a binary data file and stored to the queue by the next medium priority loop. Possible data reading irregularities are thus eliminated by the data reservoir in the queue. Third low priority loop is used for the operator commands.

Figure 4. The real-time application block diagram.

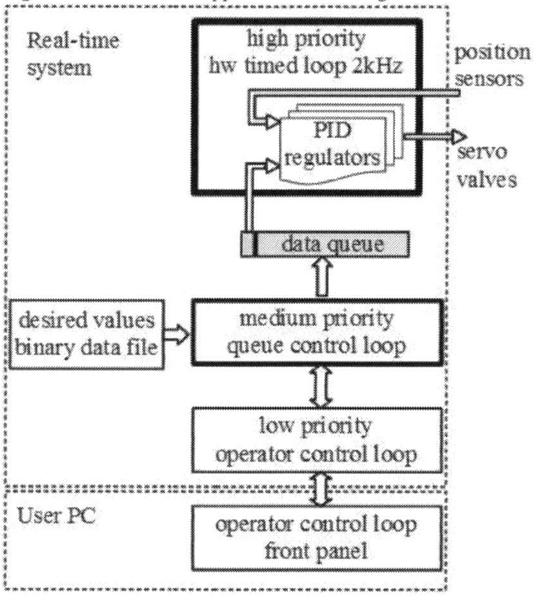

The operator control loop front panel is displayed as the user interface on the user PC. User-friendly hexapod control is thus possible. Communication between this front panel and RT application is ensured by the used real time project.

Figure 5. User interface window.

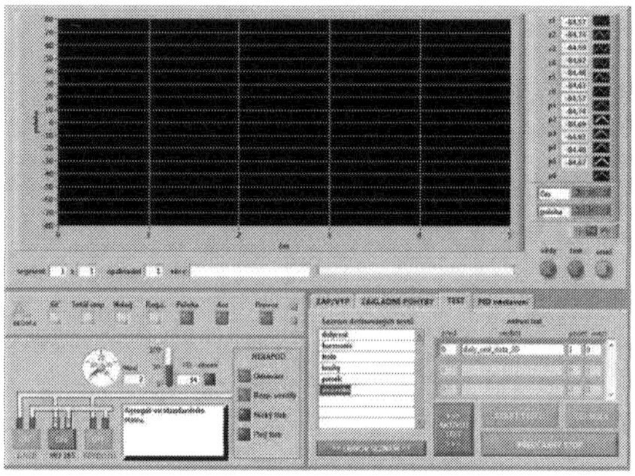

All control and display elements are concentrated in this user interface window so the summary status overview and the device control are very simple.

3.2. Real-time application optimization

Execution time of all instructions into the HW-timed loop has to be less than its timing. So the number and type of instructions have to be optimized for this condition fulfillment. This is a lengthy testing process in time optimal sequences selecting. The optimization results and

DOI: 10.1051/matecconf/20178901010

the application block diagram are shown in the Figure 6. HW-timed loop has approximately thirty percent reserve and the CPU is used on 76%. These are good results for the smooth application running.

Figure 6. The real-time application optimization result.

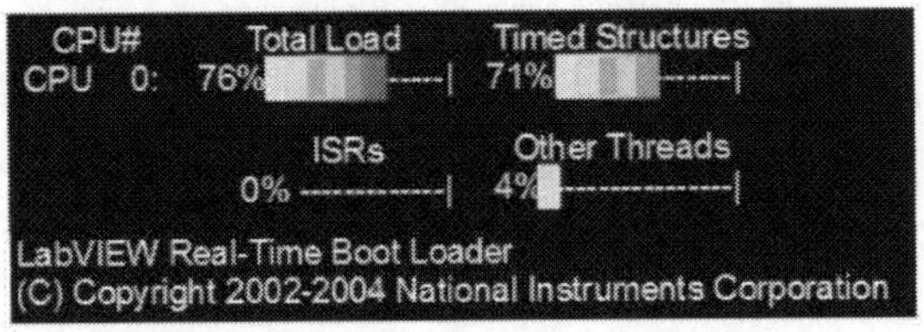

4 Hexapod kinematic calculation

Timing analysis results show that the real-time hexapod kinematics calculation is impossible in the current CPU. Only the special binary data file has to be used as the desired values source for position control.

Hexapod kinematic equations were created using its model. These equations are used to calculate the desired values binary file.

4.1. Hexapod kinematic model

The hexapod kinematic model was created using Autodesk Inventor software [3]. This model operates in the inverse mode. Central point general 3D motion (three-axis movement and three-axis rotation) is recalculated for six engines movement.

Figure 7. The hexapod kinematic model.

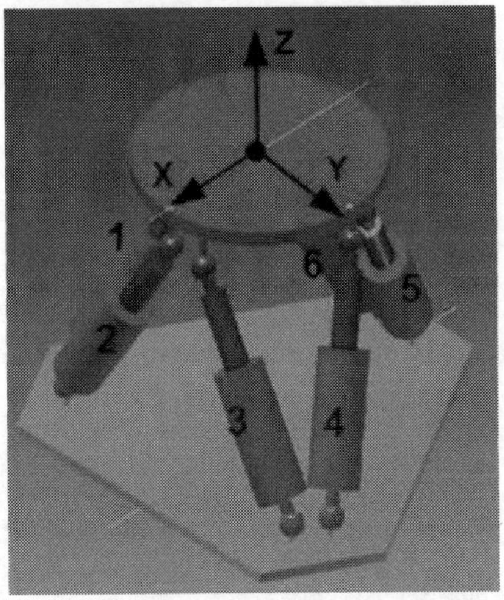

Because the calculation time was very long the conversion equations for each simple motion were created using this kinematic model. Conversion equations for the x direction are:

$$p_{eng1,4} = 0{,}0004x^2 - 0{,}4309x \quad (1)$$

$$p_{eng2,3} = 0{,}0005x^2 - 0{,}0509x \quad (2)$$

$$p_{eng5,6} = 0{,}0004x^2 + 0{,}4819x \quad (3)$$

Their graphical representation is shown in the next figure.

Figure 8. The conversion equations graphical representation.

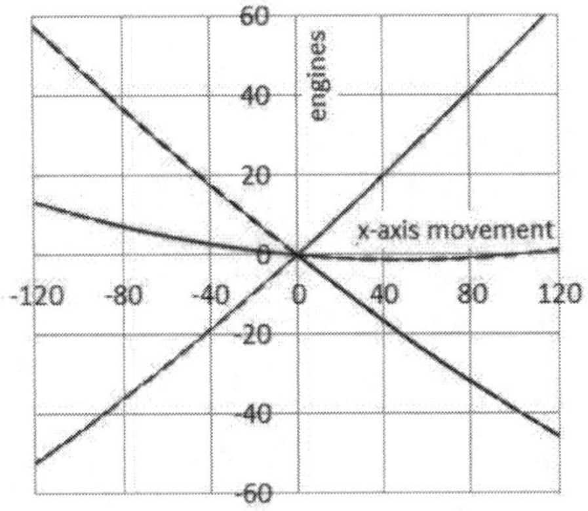

Similar quadratic equations apply to the other directions of the movement. The transformation matrix was created from all these equations and it was then implemented directly to the hexapod control software. The pre-processing application with this matrix was added to the real-time project.

4.2. Pre-processing application

Only the special binary data file has to be used for real-time hexapod position control. The desired values binary file is created before start of the hexapod movement by this pre-processing application.

Figure 9. The pre-processing application block diagram.

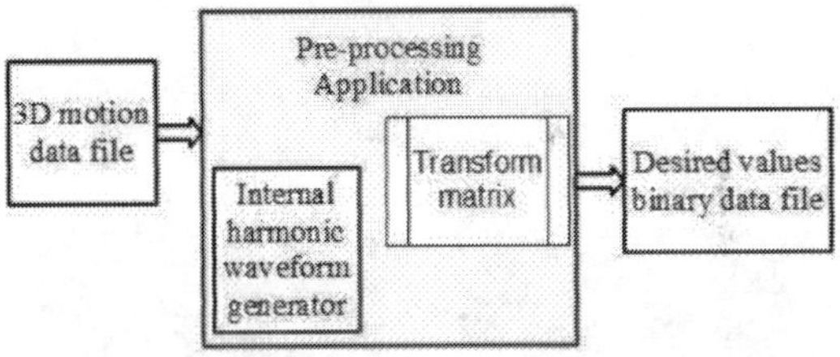

The hexapod central point movement waveforms file (its three-axis movement and three-axis rotation) is this application input. The six hexapod engines position waveforms are calculated by the transformation matrix and these waveforms are saved to the special binary format file. The binary file is then used as the reference position source during the real-time hexapod control.

This method is suitable for the real movement data (measured vibration reproduction) but for the simple movement generation is too difficult (an input data file is required). So the internal harmonic waveform generator was added to the pre-processing application and simple harmonic motion in all axes and their combination can be created directly in this application. Blocks of generated movements (including movements from the input files) can be assigned into the control sequence. Very long and varied hexapod movement sequence can be created. This application user window is shown in the next figure.

DOI: 10.1051/matecconf/20178901010

Figure 10. The pre-processing user window.

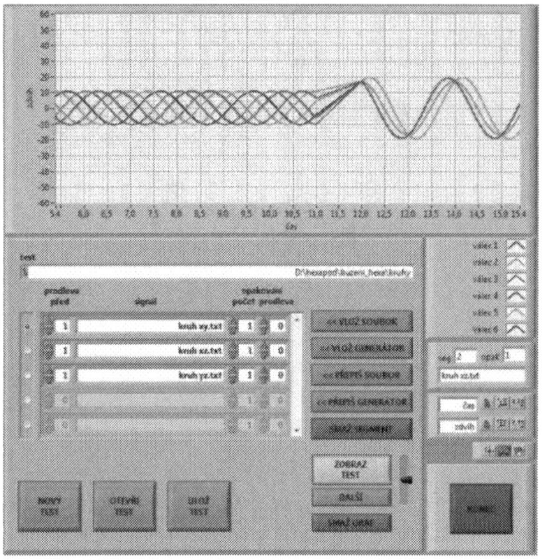

The pre-processing application is not used for real-time hexapod control so it is allocated to the user PC. Generated binary files are then moved to the RT system before hexapod motion start. This application is completely independent and it can be installed on any PC for the motion sequences generating, too.

5 Hexapod movement verification

Finally the hexapod movement is formed by relatively complicated procedure. Therefore, all component correct function had to be verified. The measuring cube was mounted on the hexapod and its actual position was measured by the noncontact laser position sensors.

Figure 11. The movement verification principle.

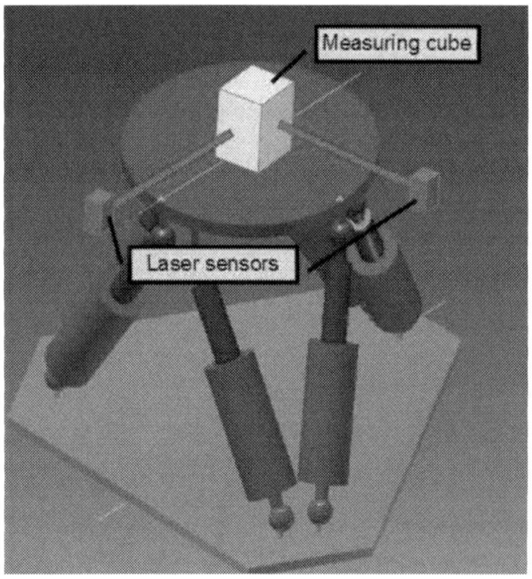

The laser position sensors were placed in different directions and the hexapod basic movements and their combinations were running. Signals from sensors were measured by the independent measuring system and then compared with the desired motion trajectory. The example of results for the combination of two phase-shifted harmonic signals in the XY plane (a circular motion in the XY plane) is shown in the next figure.

Figure 12. The verification results example.

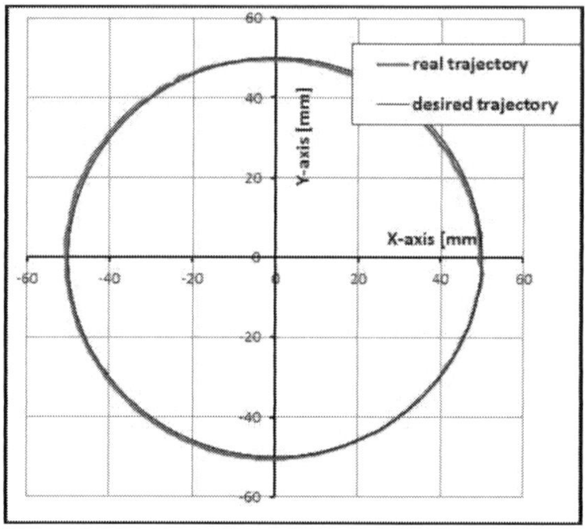

The measurements showed that the maximum deviation of the real and desired trajectory is about 1% of the hexapod movement range. It implies that the calculation algorithm and the control system operate correctly.

6 Conclusion

Conducted tests indicated that the control system and software were created correctly. The hexapod kinematic model and conversion equations well describe its kinematics and the hexapod reproduces the desired trajectory with sufficient precision. The hexapod can be used as a standard testing device.

References

1. Information on http://www.ni.com
2. A. Lufinka, Innovation of the Control System for Hydraulic Power Units. In: Modern Methods of Construction Design: Springer, 2013. pp. 477-483. ISBN: 978-3-319-05202-1
3. R. Martonka, V. Fliegel, Hexapod – the Platform with 6DOF. In: Modern Methods of Construction Design: Springer, 2013. pp. 133-138. ISBN: 978-3-319-05202-1

 DOI: 10.1051/matecconf/20178901011

Using Contactless Scanners for Quality Inspection

Radomír Mendřický[1, a]

[1]Technical University of Liberec, Department of Manufacturing Systems and Automation, Faculty of Mechanical Engineering, Studentská 2, 461 17 Liberec 1, Czech Republic

Abstract. The article presents the research of use of modern 3D measurement contactless methods for quality inspection in automotive industry production. The experience with measuring functional assemblies and parts of complex shapes as well as advantages of optical measurement methods are shown on practical research example, whose aim was to find effective procedures and methods of obtaining 3-dimensional high-definition models of measured components. The obtained models were then subjected to inspection of their dimensional and shape accuracy, which was performed by means of comparison with the nominal CAD model, as well as to analyses of the assembly functionality by searching or collision situations of the movable parts of this mechanism.

1 Introduction

Inspection of the components is currently being performed. Such inspection consists of measuring dimensional and shape accuracy, mostly by means of conventional measurement methods – for example by contact method using coordinate-measuring machine [1]. Although such measurement is very accurate, it offers only constrained amount of discrete information, usually limited to measurement and comparison of prescribed dimensions and tolerances of shape and position according to the drawing. Therefore, an assembly difficulties or collision situations of more complex assemblies and mechanisms may occur despite the fact that a part was inspected by means of the aforementioned method and is usually compliant with the prescribed dimensions.

2 Objective of the Research

The subject matter of our research was to verify and practically apply modern methods of product quality inspection by means of 3D optical and contactless digitisation. The contactless measurement methods are, as the latest research shows, becoming increasingly sophisticated and despite its lower accuracy, these methods provides more objective results than contact measurement methods in most cases. For example: Brajlih (et al.) [2] inspected the possibilities of using ATOS II 3 scanner for complex machined part geometry inspection and concluded that the coordinate-measuring machine is not suitable for parts of certain shapes. He also showed the reachable accuracy of optical scanning method by performing scanning test of gauge blocks and more well-defined geometric shapes.

In our research, we focused on selected parts of car lock with occurrence of problems with noisiness and collision situations of certain movable parts (ratchet wheels) inside the lock. All that despite the fact that the conventional measurement methods of coordinate-measuring machines did not show any specific problems and all key dimensions were within the required tolerance values. Our goal was to perform optical digitisation of these problematic parts and obtain a high-definition point cloud – an STL format model – that will help to perform an allover inspection of the obtained model when fitting on the nominal model.

Figure 1. Car lock parts

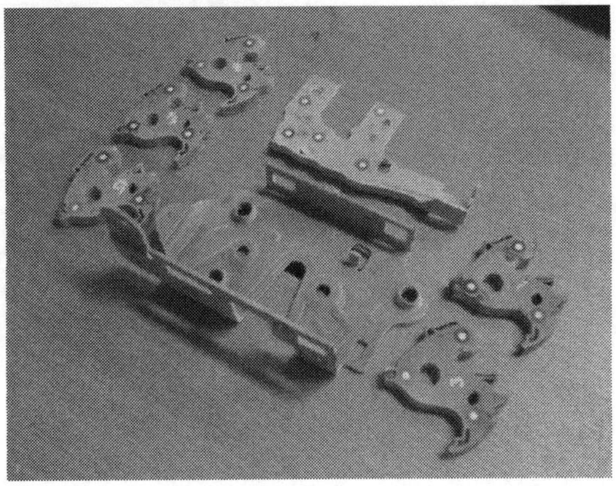

[a] Corresponding author: radomir.mendricky@tul.cz

DOI: 10.1051/matecconf/20178901011

For the purpose of reaching this goal, the digitisation method by means of a contactless optical 3D scanner was initially used on individual selected parts such as lock sheet-metal bracket, lock rear sheet, ratchet wheels of various sizes, etc. (see Figure 1). Later, the method was applied to more parts such as lock plastic body or individual pins.

In the next phase, the whole lock assembly was digitised with the optical 3D scanner, while the 3D scanning was performed for various position of the ratchet wheels (open state, locked in first position, locked in second position) – (see Figure 2).

Figure 2. Assembly of the car lock

3 Methods and equipment used:

We chose to digitise the parts with optical contactless 3D scanner manufactured by GOM – Atos II 400 (see Figure 3) fitted with lens with measurement volume of 250 x 200 x 200 mm. Before the measurement was initiated, the system was calibrated by means of standardised calibration etalon – „Calibration Panel“ (etalon approved by Decom Prüflabor laboratory).

Figure 3. ATOS optical scanner and the definition of the measurement volume [3]

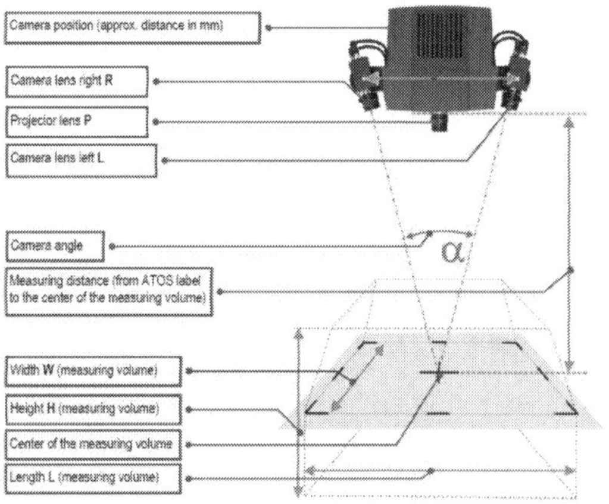

It was necessary to properly locate so called reference points on the measured part prior to initiation of the 3D scanning process. Those are contrast points and their size is given by the measurement lens used. For the measurement volume of 250 mm, points of nominal diameter of 3 mm were used (see Figure 4 – top). The points are used not only for automatic connection of partial scans to make a whole, but also due to necessity to perform the digitisation from both sides, which requires rotating the part in the process. In order to improve optical surface properties of the model, an anti-reflective chalk coating was applied on the part (see Fig. 4 – bottom). The coating is usually necessary to apply in case of products with poor optical properties in order to prevent undesirable reflections. The problematic materials are often transparent, glossy (machined metal) or even black.

Figure 4. Positioning of the reference points (top), application of anti-reflection coating (bottom)

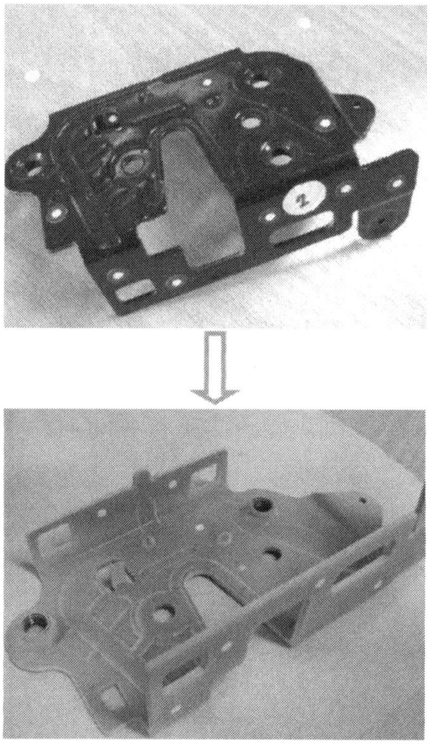

Both the assembly and individual sub-parts were attached to a GOM rotary table during the scanning process for the purpose of easier manipulation and approximately 30 to 50 scans were performed depending on the surface contours and part complexity (see Fig. 5). These created series of measurements were then connected by means of the reference points and formed a whole, which was used to calculate an optimised polygonal high-resolution polygon mesh (see Fig. 6).

Figure 5. Scanner camera images forming 3D view

Figure 6. Detail of the polygon mesh

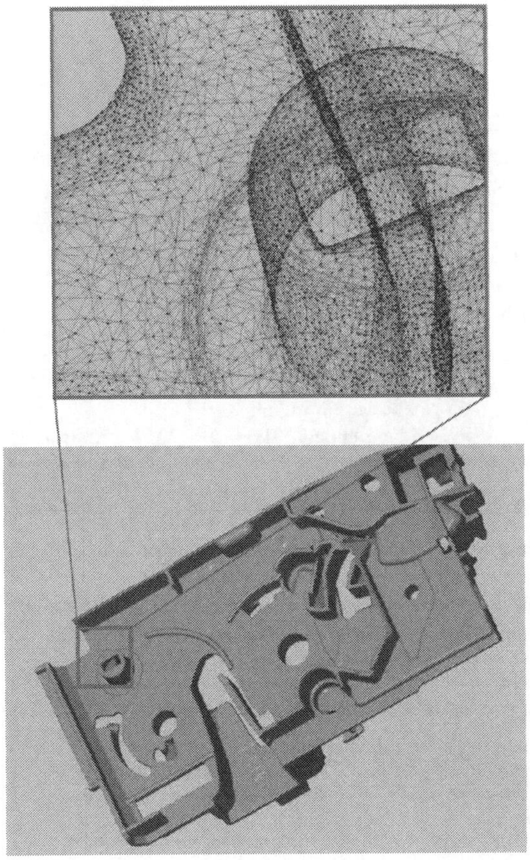

3 Research Results

Models created by means of 3D optical digitisation (see examples in Fig. 7 – scanned assembly model and Fig. 8 – digitised parts after disassembly) may be considered as actual models, while their dimensions and shape correspond to the real parts with accuracy (determined by the scanner manufacturer) – given by the measurement volume – ca. 0.03 mm, which is sufficient accuracy for this type of application.

Figure 7. STL model of the scanned assembly – ratchet wheels in closed position

Figure 8. Digitised disassembled parts

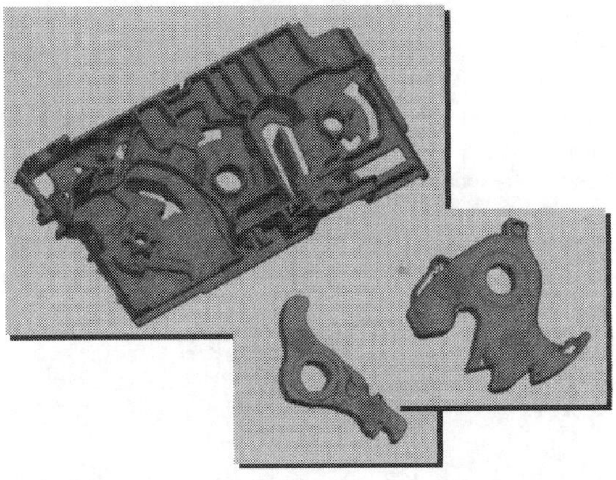

Real products – models in digitised form were subjected to dimensional and shape analysis and helped to find problematic spots in terms of the assembly. The following Figure 9 shows example of fitting an actual model to a nominal model by means of the BestFit method. When using this method, the actual model is transformed in Cartesian coordinate system in respect to the nominal model in a manner, where the total of squared deviations of all measured points on parts (so called point cloud) is as low as possible. This fitting method was used as a primary one for calculation of deviation of the real part (measured) from the nominal one (CAD).

Figure 9. The real model (grey) fitted to the shape of the nominal model (blue) using the BestFit method

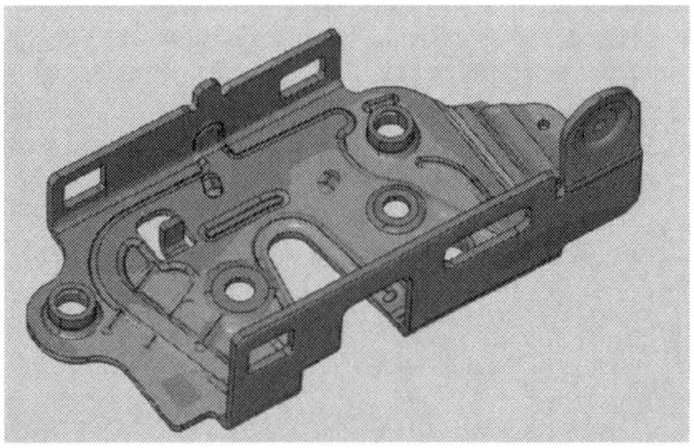

The size of normal deviations was represented for all the inspected parts and the whole assembly by so called colour deviation map. Example of colour deviation map for rear sheet is shown in the following Fig. 10. The colours in the picture demonstrate the deformation of the part, where there is too much material (warm shades – e.g. red) or vice versa, where the measured part is negatively deformed in respect of the CAD model (cold shades – e.g. blue). In addition to the allover colour map, inspection cuts may also be created. These cuts are usually created in the most problematic parts. Both, the colour maps and cuts, may be generated for various fitting to the CAD model; except the so called Overall BestFit to shape, it is usually more appropriate to use local fitting of sub-parts (so called Local BestFit) or RPS fitting (Reference Point System), often used in automotive industry.

Figure 10. Normal deviations colour map of the measured part from the CAD model (fitted by means of BestFit method)

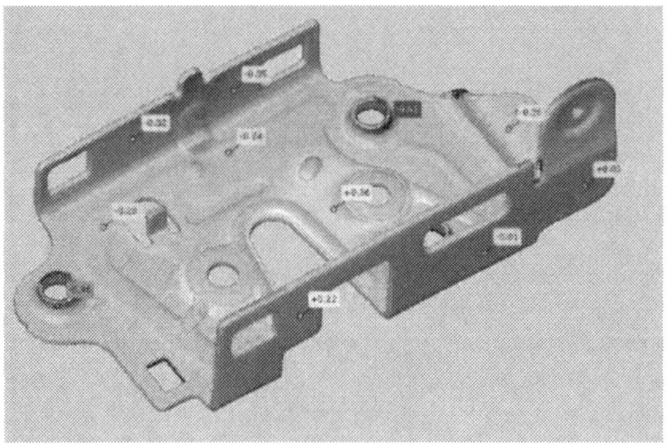

4 Summary

For the purpose of car lock quality inspection, a contactless optical method of scanning 3D objects by means of ATOS II 400 optical scanner was designed and successfully used. The models of real parts obtained with this digitisation method were reaching a density of appx. 3,000 points per 1 cm^2, thus allowing performance of analyses with high precision and detail. Creating complex high-definition models of parts contributed to seeking new possibilities of production quality control not only for individual parts, but for functional assemblies as well. Digitised parts transformed to STL format were used for quality inspection when comparing to the CAD nominal template, functionality of the whole assembly, as well as for searching for collision situation of the movable parts of mechanism. The research contributed to solving problems with noisiness of the lock and to improving functional quality of these products. The test also proved that high-tech optical scanning systems may be reliably used even for complex geometry inspection of complex parts and assemblies. Referring to the current and anticipated future development of optical 3D scanning and to the increasing accuracy of these devices, these systems will apparently expand to other fields of inspection as an alternative to conventional measurement methods, thus contributing to increase of productivity and production quality, as well as to increase of product innovation and development speed.

Acknowledgement

The research reported in this paper was supported by targeted support for specific university research within the student grant competition TUL (Project 21010 - Complex optimization of manufacturing systems and processes 2).

References

1. S. T. Chang, W. C. Lin, T. M. Huang, M. Y. Hsu, P. H. Huang, Measurement of Radius of Curvature by Coordinate Measurement Machine, Applied Mechanics and Materials, Vols 284-287, pp. 488-492, (2013)
2. T. Brajlih, et al. Possibilities of Using Three-Dimensional Optical Scanning in Complex Geometrical Inspection. Strojniški vestnik - Journal of Mechanical Engineering [online]. (2011), roč. 57, č. 11, s. 826-833 [cit. 01/04/2015]. ISSN 00392480. Information on: http://en.sv-jme.eu/data/upload/2011/11/05_2010_152_Brajlih_05.pdf
3. GOM MBH. ATOS Compact Scan: User manual - Hardware. Braunschweig, Germany, (2012)
4. R. Mendřický, Use of Optical Methods for Dimensional Analysis of Multi-Part Moulds. In. Applied Mechanics and Materials, Vol. 474, Trans Tech Publications, Switzerland, (2014), pp 387-392, ISSN 1660-9336
5. S. Zhang, Handbook of 3D Machine Vision: Optical Metrology and Imaging (Series in Optics and Optoelectronics), Publisher: CRC Press, (2013), ISBN: 978-1439872192

 DOI: 10.1051/matecconf/20178901012

Stress of dynamic loaded protection cab ROPS of truck

Ladislav Ševčík[1], Ludvík Prášil [1]

[1]Technical University of Liberec, Dept. of Machine design, Studentská 2, 461 17 Liberec 1, Czech Rep.

Abstract. The paper describes the design geometry, calculate deformation, deformation energy and stress safety of extended hydraulically lifted ROPS frame lorry TATRA by FEM according to Australian Standard AS 2294.2-1997 Earth-moving machinery-Protective Structures, Part 2: Laboratory tests and performance requirements for roll-over protective structures and by international Standard ISO 3471: 2008 (E) for the proposed design solution.

1 Introduction

The aim of the study was to investigate the protective structure of the cab of a truck TATRA 8x8 to cater for Australian standards. Suggest a change ROPS a proposal verify calculations deformation, stress and energy loss when objects falling on the car cabin. The car is built into mining operations and large excavators an object may fall on a passing truck.

2 The values specified for the design and calculations ROPS

The input data and documents geometry was shaped contours cabin, which was carried out according to the design of the safety of prolonged ROPS frame see Figure 1 front view of the cab cover ROPS and Figure 2 lateral view of the cab cover ROPS with off-road truck TATRA 8x8. It was made of closed profiles 80x80x8. Design was made in program AutoDESK INVERTOR. Furthermore, the maximum values were given static forces onerous safety ROPS acting in a direction lateral, longitudinal and vertical direction and in the lateral effects of kinetic energy, viz. the table 1, calculated in accordance with Australian Standard AS 2294.2-1997 and international standards ISO 3471: 2008 (E), and the material elements of design was use the unalloyed structural steel - designation by CSN 11 523 (or ST 52).
The main dimensions of safety frame, the size of the forces and their sphere of activity are the same for all admissible solutions of the protective frame. Location of protective beams is given by norm. The only difference is the size of the square or rectangular cross-section thin-walled closed profiles. Thin square profiles can be reinforced by flat rods with dimensions 80x12x1000mm.
For the following design and calculation must be observed.

Figure 1. Front view of the cab cover ROPS. [2]

Figure 2. Lateral view of the cab cover ROPS [2].

Orientation of the model in a rectangular coordinate system of safety frame:
x axis - the direction of the lateral (lateral) forces

y axis - the direction of the longitudinal force (against cavalry)
z axis - direction vertical forces
In Table 1 are displayed the value of forces acting in different directions of the coordinate system.

Table 1. The value of forces acting in different directions of the coordinate system.

Direction	*x* (lateral)	*y* (longitudinal)	*z* (vertical)
Force*)	$F_2 \equiv F_x =$ +105 778 N	$F_3 \equiv F_y =$ -84 630 N	$F_1 \equiv F_z =$ -235 320N
Energy*)	$U_2 \equiv U_x =$ 18 839 J	-	-

*) Values of all forces and energies in the direction of the lateral were calculated weight of the machine M = 12,000 kg according to Australian Standard AS 2294.2-1997, Tab. 1 - The equation for calculating the forces and energy (Force and energy equations), line 7) Rigid frame machines -only for ROPS (Rigid frame dumpers - ROPS only option) and the International Standard ISO 3471: 2008 (E).

The design of the protective frame allows its adjustment in two positions. At work, movement position the frame is secured by two pins whose movement is derived from the linear hydraulic motors see. in Figure 3.

Figure 3. Movement position the frame of the cab cover ROPS.

For any corrections and adjustments to the car, changing unlocks the pins and turning to see the service position see Figure 4.

Figure 4. Lateral view of the cab cover ROPS

The boundary conditions for the calculation are shown in Figure 4. The figure shows the forces acting according to the standard. Cross member is only auxiliary and represents a testing device that distributes the load according to all stringers evenly. Protection ROBS is rigidly fixed to the frame of TATRA truck.

Figure 4. Raised position the frame of the cab cover ROPS.

3 Calculation of strains and stresses

Assuming perfect penetration welds the parts connected to the entire construction simplified so that the load is transferred only sections of the work pieces structure, ie. it is not considered material located outside the weld these sections. This simplification is to be done and it can be assumed that significantly affect the results of calculations. FEM calculations are performed for the material constants: modulus of elasticity E = 200 GPa and Poisson's ratio $\mu = 0.3$.

Strains and stresses were calculated for lonely forces amid distributors load (distributors) located:

- For lateral loads in a direction (+ x) (lateral load) on the side beam of the frame, the distributor has a length of

66.6% of the distance (L) between the outer front and rear lateral of the upper rectangular frame ROPS and is positioned according to a distance of 1 / 3L from the rear lateral of the upper rectangular frame ROPS.

For longitudinal load in the direction of travel (s) (Longitudinal load) in the middle on the rear frame beam width distributor is max. 80% of the distance (W) between the outer left and right sides of the upper rectangular frame ROPS

For the vertical load in the axial direction (z) (vertical load) in the middle of the vertical load beam (not part of the frame ROPS, see. Figure 4) located midway between the front and rear outer side of the upper rectangular frame ROPS.

On color maps are shown the calculated values of deformation (displacement) and equivalent (reduced) stress by HMH, Figure 5-10. On the color scale is assigned to each color range of strains and stresses. Places where the stress exceeds 500 MPa is colored deep red. These are eg. space boundary conditions of load and sudden transitions. In terms of stress are the most unfavorable values at the connection profiles and corner elements, where there (without considering welds) a sudden change in shape. In these places it is a local (local) peak stresses that causes plastic deformation in a small volume of material and will not affect the resistance to static cross-section or a single load. In these places it is necessary to read the values calculated stress at a reasonable distance from their connections (about two to three times the thickness of the respective structural member).

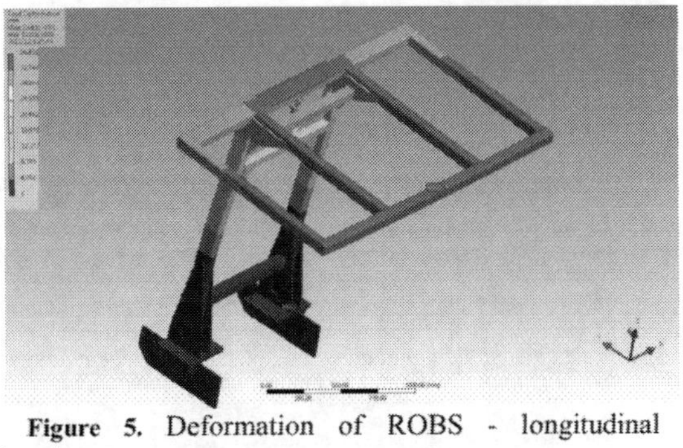

Figure 5. Deformation of ROBS - longitudinal direction.

Figure 6. Stress of ROBS - longitudinal direction.

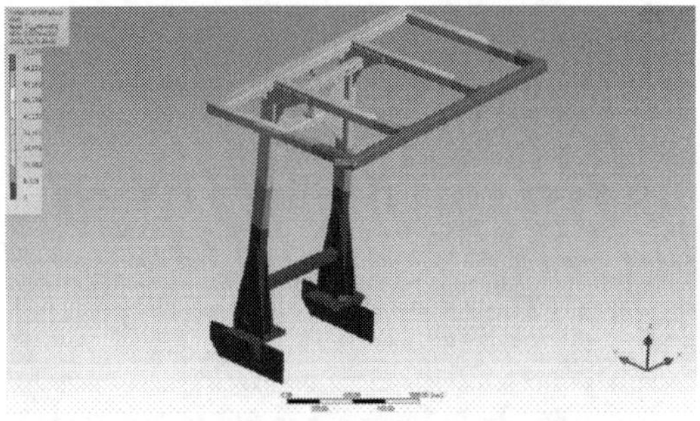

Figure 7. Deformation of ROBS - lateral direction.

Figure 8. Stress of ROBS - lateral direction.

Figure 9. Deformation of ROBS - vertical direction.

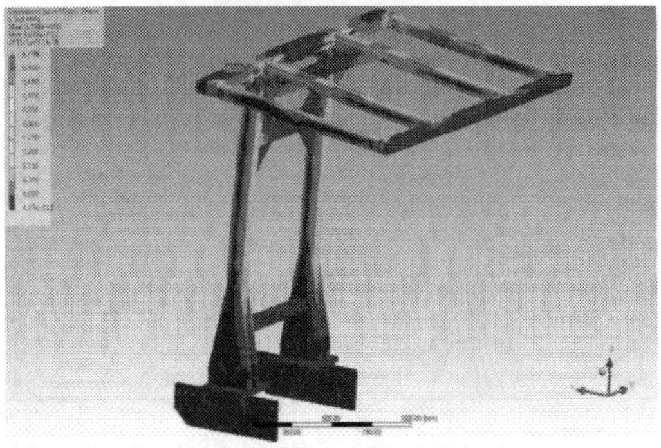

Figure 10. Stress of ROBS - vertical direction.

Evaluation of energy absorption will be considered according to the mechanical energy expended on its deformation in bend – mechanical energy is determined by the following equation [1]:

$$U = \int_{x1}^{x2} F(x)dx \quad (1)$$

In the preceding formula represents:
U - mechanical energy [J]
F(x) - function of the force-deformation of the ROPS [N]
x - deformation of test [m]

The energy calculation was performed on elasto-plastic model, Figure 11,12.

Figure 11. Deformation of ROBS - lateral direction, elastoplastic model.

Figure 12. Stress of ROBS - lateral direction, elastoplastic model.

3 Conclusion

The resulting of stress and deformation for all variants of the loading are shown in the following Table 2. The modified structure of the safety bracket (third variant) achieved a reduction of the overall deformation in a lateral load force and lateral energy, viz. in Table 2.

Table 2 .: Deformation and stress ROPS in spatial transformation structures.

	Application of the force and energy*)	Deformation of variant th.3 [mm]	Stress of variant th.8 [MPa]
Max. deformace /mm/	vertical	36,40	4100
	lateral	72,54	5700
	longitudinal	62,82	6800
Deformace **) /mm/	lateral*)	573	4710

*) After reaching the yield strength of said calculation steps in the elasto-plastic deformation of 10x lower elastic modulus value, i.e. $E_{e\text{-}pl} = 20$ GPa

**) In terms of deformation is in the middle of distributors, ie. when a force corresponding to the calculated value of lateral energy.

The sponsor performed on the recommendation of the investigator changes and process third structural variant of the frame. The calculation results show that by modifying the original design (Option 3) reinforcement flange square thin-walled profile 80x80x6mm flat bar dimensions 12x80x1000mm achieved a reduction of distortions in the vertical direction by 28.3% and in the lateral direction by 18%.

By reinforcing the structure of the safety frame was also achieved a reduction of the equivalent stresses in the most loaded cross-sections of the structure. This state is already apparent by simply comparing images that show stress in the same manner load.

In terms of stress are the most unfavorable values particularly at the connection profiles and corner elements, where there is (without considering welds) a sudden change in shape. In these places increased local operating voltage that causes plastic deformation in a small volume of material cross section, so there is a failure in the static or single dose loading. In these places it is necessary to read the values calculated stresses within a reasonable distance of the connection profiles (about two to three times the thickness of the respective structural member).

Based on measured values of deformation and overall stress of prolonged construction safety, hydraulically lifted is ROPS frame. It is clear that not considered at maximum load, according to the standards laid down by AS 2294.2 - 1997 and ISO 3741: 2008 (E) the total destruction of the safety bars.

Figure 13. Protective equipment installation ROPS cab truck.

References

1. Malik L., Medvecky S., Casti a mechanismy strojov; TU Zilina, January 2004.
2. Information on http://www.tatra.cz/o-spolecnosti/tisk-a-media/aktualne/tatra-phoenix-aplikace-pro-tezarsky-prumysl/

DOI: 10.1051/matecconf/20178901013

The Effect of Flywheel Unbalance on Gear Noise in the Hydraulic Power Plant Turbo-Generator

Elias Tomeh

Technical University of Liberec, Department of Vehicles and Engines, Studentská 2, 461 17 Liberec, Czech Republic

Abstract: The Effect of Flywheel Unbalance on Gear Noise in the Hydraulic Power Plant Turbo-Generator. Hydraulic power plants are systems that produce electrical energy with high investment costs. In order to fulfil their goals, investments should create conditions for a safe production of energy in a long lasting and reliable way, and with the required power and quality. These goals are possible to reach by an optional control process linked to a systematic monitoring of the operating machinery state, using the method of vibration diagnostics. Lately, there has been an increase of noise level in the hydraulic power plants.

1 Introduction

Technical diagnostics of machines and equipment based on vibration measurement is currently applied in all industry sectors with substantial economic effect.

Achieved results in operating application of diagnostics (reducing maintenance costs and increasing reliability of machines) confirm that the diagnostics is a necessity for the operation of modern machines and equipment. Vibration measurements, in addition to early identification of malicious damage to the components and groups of monitored systems, also objectively evaluate the quality of manufacturing, assembly and adjustment of new or repaired machines. Thanks to the periodic evaluation of measurements, inexpensive measures enabling to prevent damage can usually be designed and implemented as well.

In the technically advanced countries, a maintenance based on the actual determination of the reported system condition, is applied in hydroelectric power plants. This means that:

- Shutdown of machines will only be made in cases where it is required because of their technical condition,
- Machine components are changed only in the event of wear or damage,
- Rotor balancing is performed only when it exceeds the prescribed limits,
- Alignment adjustment of the machine is performed when the deviation is large.

Abroad, in hydroelectric power plants, the vibration measurement is used for operational condition monitoring (in addition to the measurement of other variables such as bearing temperatures). The relative vibration of shaft is measured with proximity displacement sensor and the absolute vibration at the bearing point is identified.

In accordance with international knowledge, measurements of the velocity of absolute vibration were taken on TG machine. After general maintenance in the body of the small hydroelectric power plant, an increased vibration occurred at the designated measuring points. See Figure 1.

Acceleration sensors at the mentioned points were fixed by a permanent magnet. During the measurement, the machine was operating in a steady mode with a constant generator power of 100 kW.

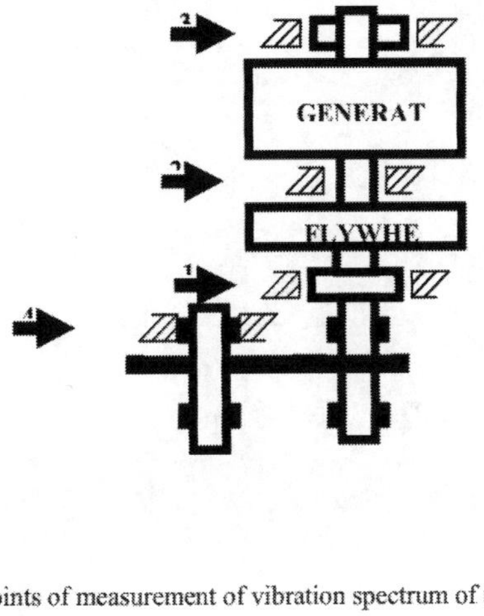

Figure 1. Points of measurement of vibration spectrum of the Turbo Generator TG

Corresponding author: elias.tomeh@tul.cz

The overall level of machine vibration are assessed according to the effective value of the vibration velocity

$$v_{ef} = \sqrt{\left(\frac{1}{T} \cdot \int_0^T v^2(t)dt\right)},$$

Where:
v(t) ... is time-dependent vibration velocity [mm.s^{-1}],
$\mathbf{v_{ef}}$ ... corresponding total velocity value [mm.s^{-1}],
T ... measuring time [s].

During the decomposition of mixed vibration into the frequency spectrum, $\mathbf{v_{ef}}$ can be determined in the relationship

$$v_{ef} = \sqrt{\sum_{i=1}^{n} v_{oefi}^2} \quad \text{[mm.s}^{-1}\text{], where}$$

$\mathbf{v_{oefi}}$ is the i-th effective vibration velocity amplitude of the i-th harmonic spectrum component.

The maximum value of $\mathbf{v_{ef}}$, measured on the important points of the machine, indicates the vibration severity U according to:

$$U = \left(x_{ef}, v_{ef}, a_{ef}\right)_{max}$$

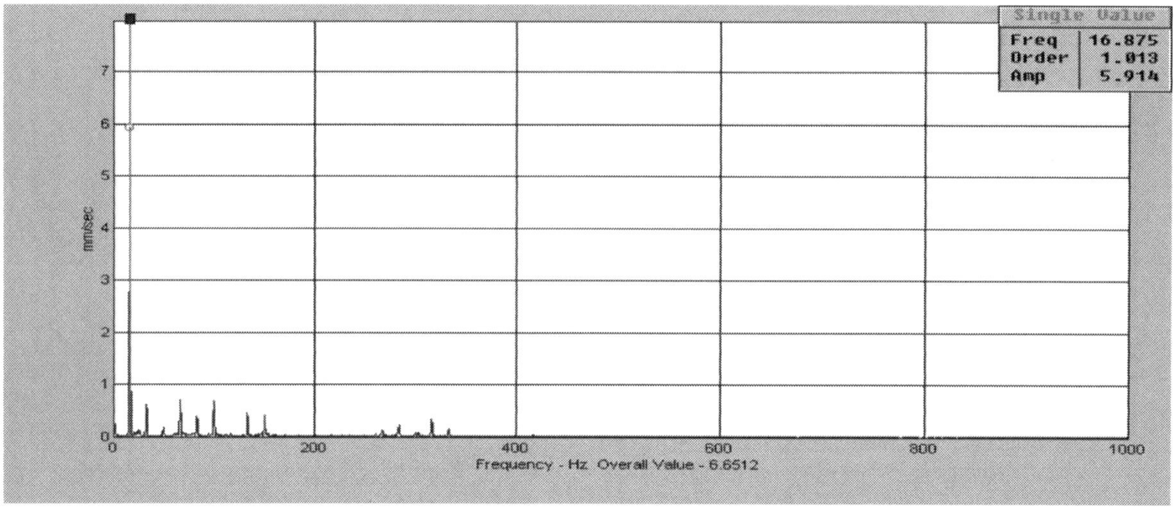

Figure 2. Vibration spectrum TG in **3H point before balancing** and with load, RPM 1012, the amplitude of the rotor frequency $V_0 = 5.91$ mm, $v_{ef} = 6{,}65$ mm.s^{-1}.

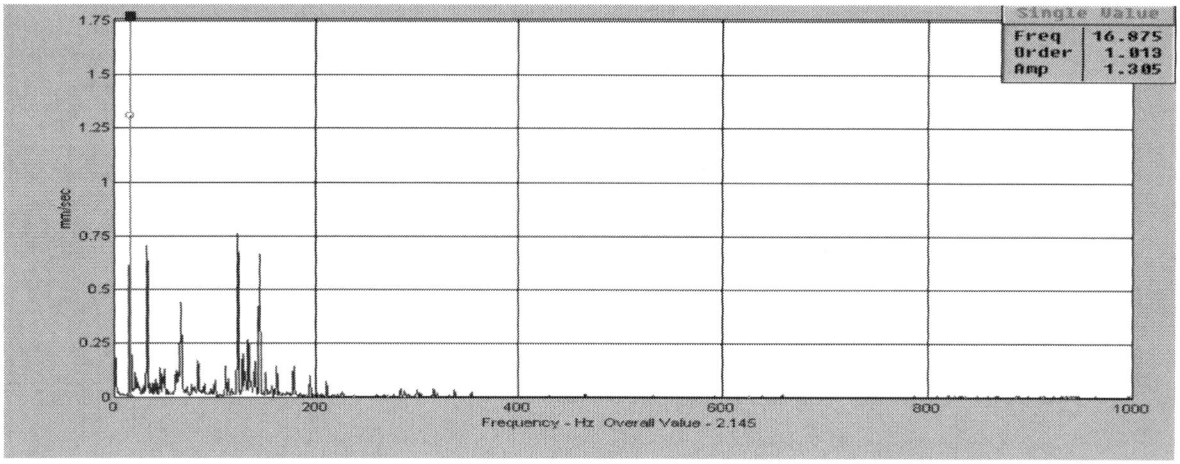

Figure 3. Vibration spectrum TG in **3H point after balancing** and with load, RPM = 1012, the amplitude of the rotor frequency $V_0 = 1.38$ mm, $v_{ef} = 2{,}14$ mm.s^{-1}.

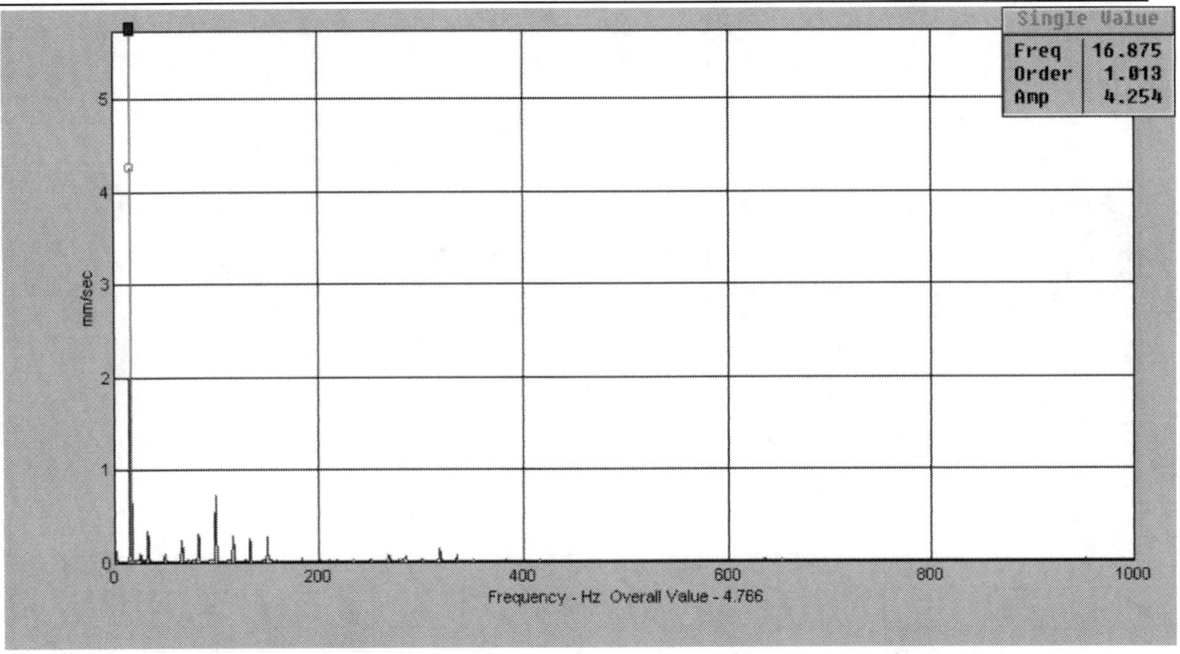

Figure 4. Vibration spectrum TG in **2H point before balancing** and with load, RPM = 1008, the amplitude of the rotor frequency $V_0 = 4.25$ mm, $v_{ef} = 4{,}766$ mm.s^{-1}.

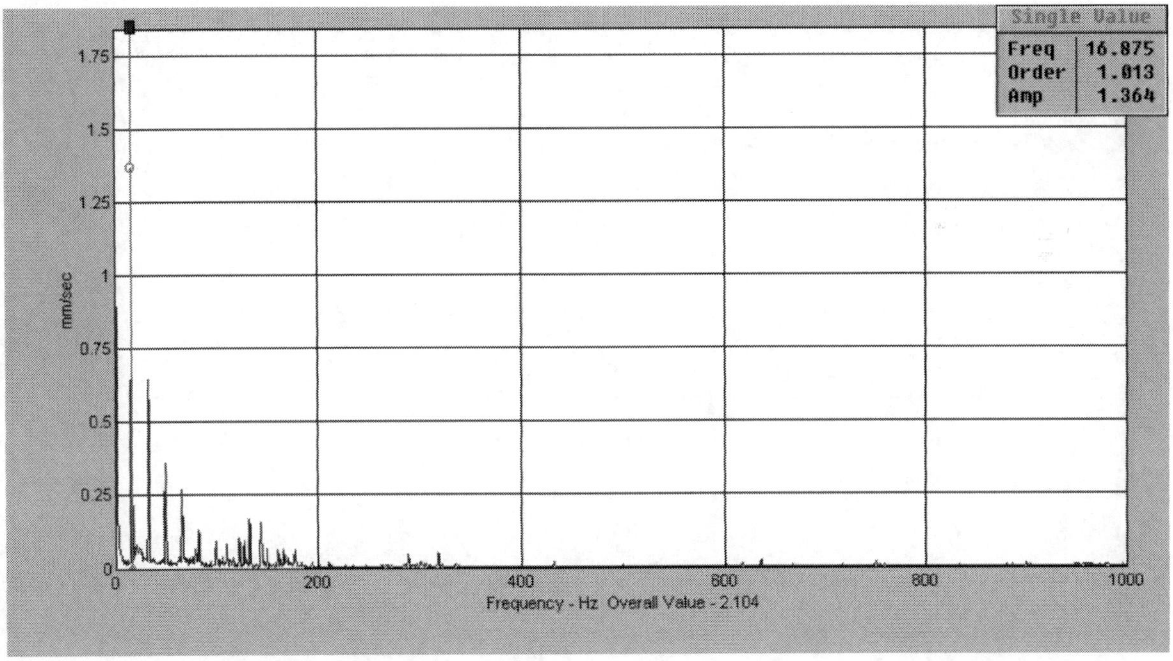

Figure 5. Vibration spectrum TG in **2H point after balancing** and with load, RPM = 1012, the amplitude of the rotor frequency $V_0 = 1.36$ mm, $v_{ef} = 2{,}1$ mm.s^{-1}.

 DOI: 10.1051/matecconf/20178901013

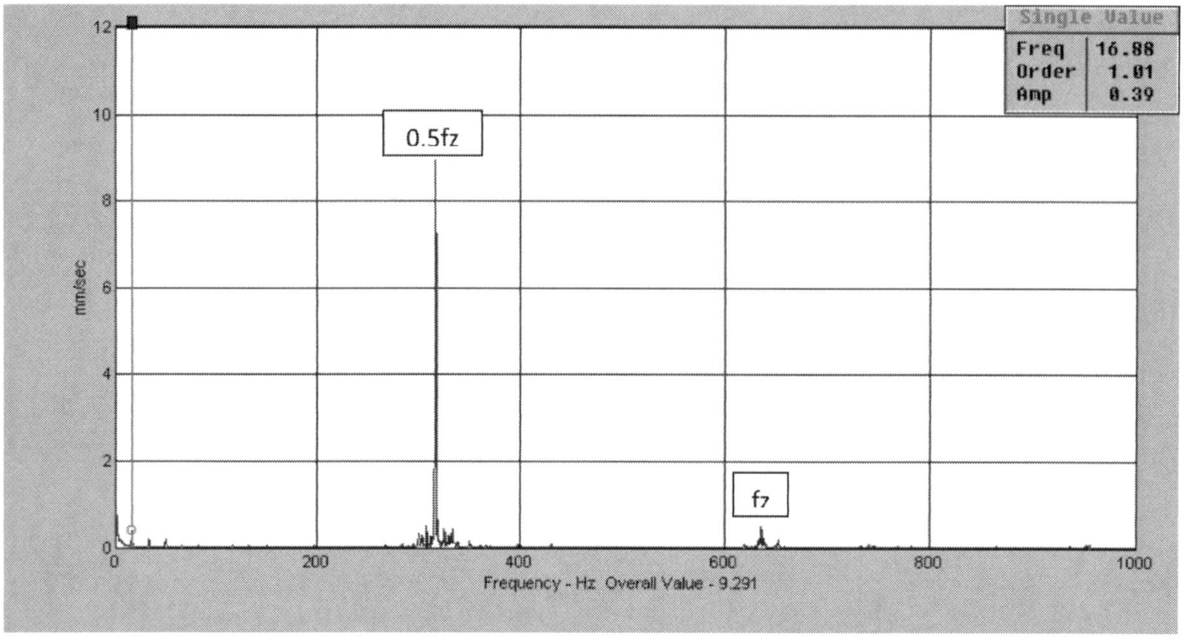

Figure 6. Vibration spectrum TG in **1H point before balancing** and with load, RPM = 1008, the amplitude of the rotor frequency V_0 = 0.39 mm, v_{ef} = 9,291 mm.s^{-1}, the amplitude of gear frequency 0.51 mm, the amplitude of gear frequency interharmonic component 0.5 f_z = 8.93 mm.

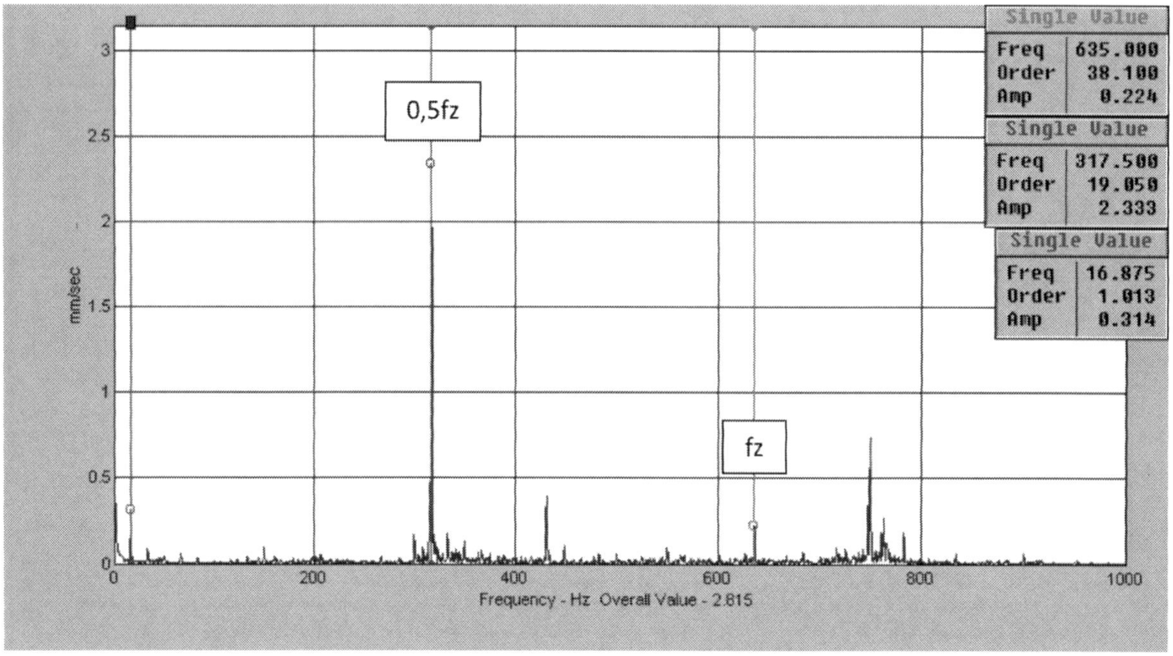

Figure 7. Vibration spectrum TG in **1H point after balancing** and with load, RPM = 1012, the amplitude of the rotor frequency V_0 = 0.3 mm, v_{ef} = 2,815 mm.s^{-1}, the amplitude of gear frequency 0.22 mm, the amplitude of gear frequency interharmonic component 0.5 f_z = 2.3 mm.

DOI: 10.1051/matecconf/20178901013

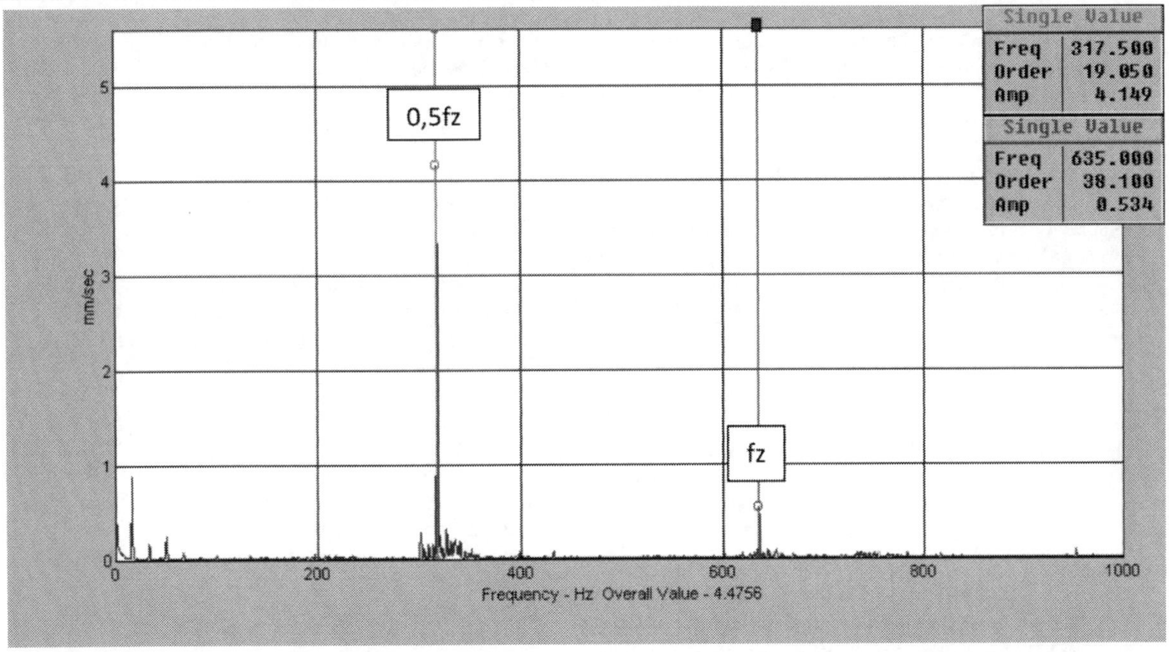

Figure 8. Vibration spectrum TG in **4H point before balancing** and with load, RPM = 1012, the amplitude of the rotor frequency $V_0 = 0.87$ mm, $v_{ef} = 4.47$ mm.s^{-1}, the amplitude of gear frequency 0.53 mm, the amplitude of gear frequency interharmonic component 0.5 $f_z = 4.14$ mm.

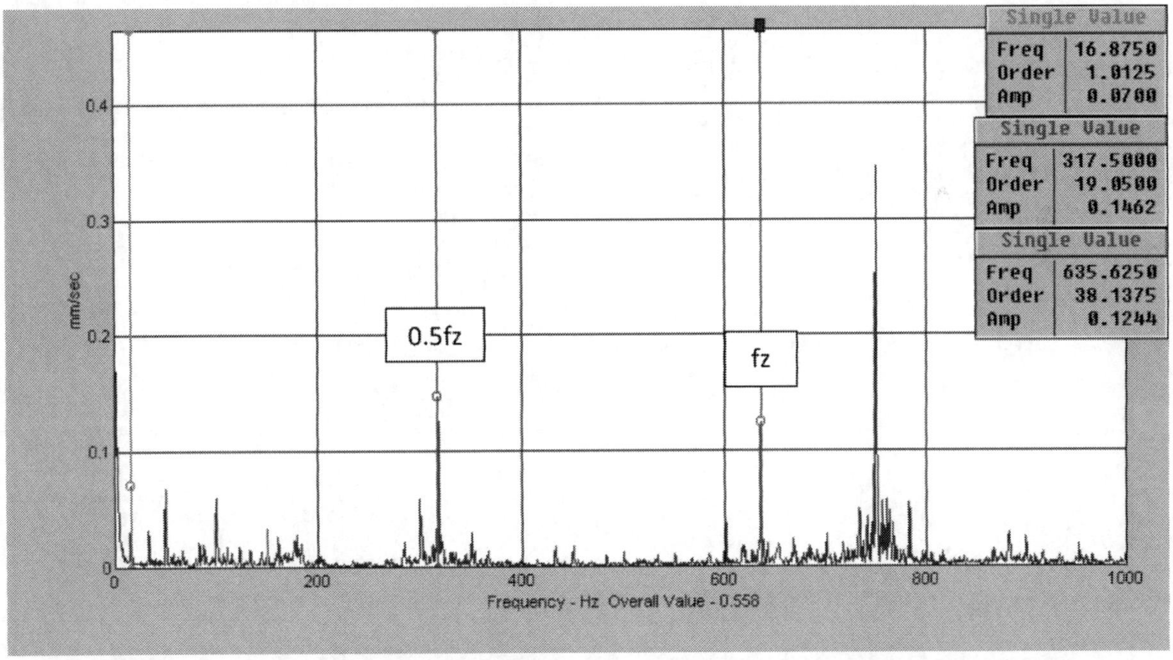

Figure 9. Vibration spectrum TG in **4H point after balancing** and with load, RPM = 1012, the amplitude of the rotor frequency $V_0 = 0.07$ mm, $v_{ef} = 0,55$ mm.s^{-1}, the amplitude of gear frequency 0.12 mm, the amplitude of gear frequency interharmonic component 0.5 $f_z = 0.14$ mm.

For TG machine, the greatest level of vibration was measured at the turbine in the horizontal direction 1H. The component of the frequency fz (gear frequency) and the component of the frequency fR (rotor frequency) determine the overall level of vibration. Because the amplitude V0 of the rotor frequency fR increases with the size of power and current, it can be said that the vibration excitation is directly related to the mechanical unbalance of the generator rotor flywheel.

The measurement demonstrated that the turbo-generator TG has at full power a quite high vibration level U = 9.29 $mm.s^{-1}$.

The analysis of the vibration measurements of the small hydroelectric power plant TG (Fig. 2, 4, 6 and 8) showed that the main cause of the excessive levels of vibration of flywheels and high noise pollution of gears is the unbalance of rotor turbine with the flywheel.

Standard ČSN ISO 10816 Vibration - is the basic document establishing general guidelines for the machines vibration measurement and evaluation. Criteria for evaluation of vibration were set especially with regard to ensuring a safe and long-term operation of machines. From the measured vibration severity, the small hydroelectric power generator TG, according to [1] can be allocated to Class II. According to article 5.3.1 of this norm, the TG machine can be classified in range C: U = 2.8 to 7.1 $mm.s^{-1}$. Machines, whose vibration is in the C range, are normally considered as unsatisfactory for long-term continuous operation. Generally, the machines can operate under such conditions until we find the possibility of correction. It has been suggested to carry out an operational dynamic balancing of the machine flywheel by adding the correction weight in balancing plan 1 (131.7 g, phase angle 122°) and in balancing plan 2 (64.2 g, phase angle 87.5). The TG results after balancing the machine flywheel rotor are listed in Tab. 1. and illustrated in Figure 5. and Figure 7.

The overall machine condition can be according to [1] and the measured overall vibration level evaluated as acceptable level of vibration. In case of continuous long-term operation of this generator, the life expectancy of the machine components will probably be lower.

For comparison the measured vibration severity is included:

- U = 2.7 [mm.s-1], a large turbine AESCH Switzerland
- U = 3.7 [mm.s-1], MVE Ervěnická corridor, sets 1
- U = 3.2 [mm.s-1], MVE Ervěnická corridor, sets 2
- U = 4.2 [mm.s-1], MVE MVE Orion Liberec
- U = 9.29 [mm.s-1], MVE Josefův Důl

The numerical values of vibration severity for the purpose of acceptance should be the subject of agreement between the machine manufacturer and the customer. The purchase agreement has to include:

- Machine inspection also on the basis of vibration and noise measurement,
- Cover of costs for the measurement and elimination of possible faults (tuning) - reduction of the contract price of the machine.

EU directive for engineering requires that the measured vibration severity and the equivalent noise level be included in the documentation that accompanies the product.

Measurement and analysis of vibration (noise) are also useful in other areas of practice:

— **Development of new machines** – The measurement effectuated on manufactured machines provides an objective basis for the tuning and adjustment of machines - long life, high reliability and safety.

— **Machine vendor selection** – The name of the manufacturer, the price, service conditions and promotional data may not be decisive.

Table 1. Measured value of the total vibration level [$mm.s^{-1}$] and significant components of the spectrum V0 with load Pg = 100 kW

	Point of measurement	**TG** **Generator power Pg=100 kW**			
		Overall [$mm . s^{-1}$]	V_0 (f_R) [$mm . s^{-1}$]	V_0 (f_z) [$mm . s^{-1}$]	V_0 (0,5f_z) [$mm . s^{-1}$]
Before balancing	1H	9,29	0,39	0,51	8,93
	2H	4,76	4,25	-	-
	3H	6,65	5,91	-	-
	4H	4,47	0,87	0,53	4,14
After balancing	1H	2,81	0,30	0,22	2,3
	2H	2,10	1,36	-	-
	3H	2,14	1,38	-	-
	4H	0,55	0,07	0,12	0,14

 DOI: 10.1051/matecconf/20178901013

2 Acquired knowledge

The dynamic tuning of TG flywheel resulted in a significant decrease in the machine vibration and a significant reduction of gears noise.

Due to the elimination of flywheel unbalance in points 1H, 2H, 3H and 4H, the overall level of vibration of the TG turbine decreased as well as the vibration frequencies amplitude corresponding to the rotor unbalance. At the same time, the amplitude of gear frequency and the amplitude of the interharmonic component frequencies (0.5 gear frequency) decreased. This interharmonic frequency of gears clearly shows the load on the pinion gear of the turbine wheel, due to flywheel unbalance. See Table 2 and graphs.

In this state the machine TG achieves long lifetime in continuous operation.

Table 2. The decrease in the size of the individual vibration components after balancing the TG flywheel

The decrease in size of the individual vibration components [%]				
Point of measurement	Overall [mm . s^{-1}]	V_0 (f_R) [mm . s^{-1}]	V_0 (f_z) [mm . s^{-1}]	V_0 (0,5f_z) [mm . s^{-1}]
1H	**70 [%]**	**23 [%]**	**57 [%]**	**74 [%]**
2H	**56 [%]**	**68 [%]**	-	-
3H	**68 [%]**	**77 [%]**	-	-
4H	**88 [%]**	**92 [%]**	**77 [%]**	**97 [%]**

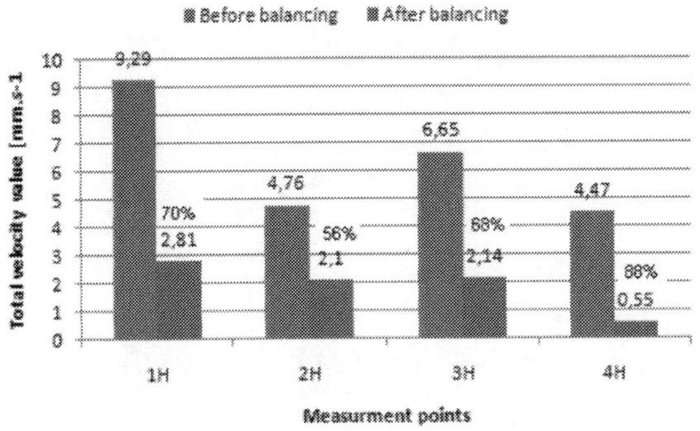

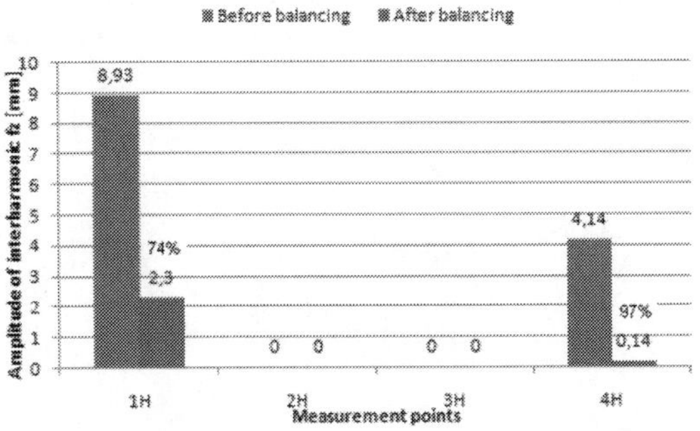

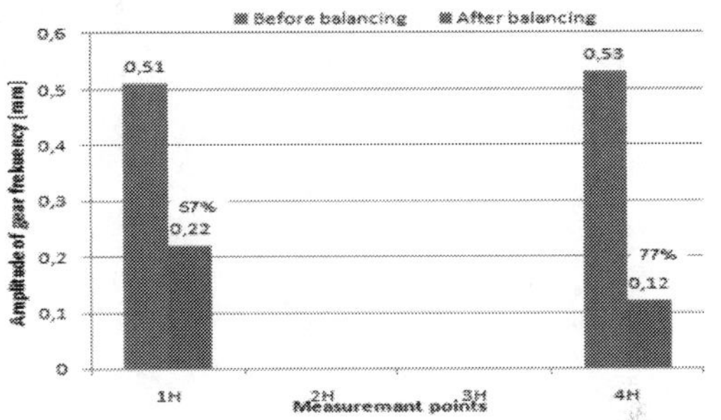

RBferencesR

1. ČSN ISO 10 816-1 Vibrace - Hodnocení vibrací strojů na základě měření na nerotujících částech. Část 1: Všeobecné směrnice.
2. Beneš, Š. – Tomeh, E.: Posouzení vibrací malé vodní elektrárny Kolora Přepeře. TU Liberec 1995.
3. Angelo, M.: Monitorování mechanického chvění strojních zařízení. Brüel & Kjaer 1987.
4. Tomeh, E.: Diagnostic Methodology of Rolling Element and Journal Bearings. /Skripta TU v Liberci/, TUL Liberec, 2007.
5. Tomeh, E.: Technická diagnostika – Vibrační diagnostika strojů a zařízení. /Skripta TU v Liberci/, TUL Liberec, 2015.

AUTHOR INDEX